T0297782

Intelligent Edge Computing for Cyber Physical Applications

Intelligent Data-Centric Systems

Intelligent Edge Computing for Cyber Physical Applications

Edited by

D. Jude Hemanth
Karunya University, Coimbatore, Tamil Nadu, India

Brij B. Gupta
International Center for AI and Cyber Security Research and Innovations, & Department of Computer Science and Information Engineering, Asia University, Taichung, Taiwan

Mohamed Elhoseny
College of Computer Information Technology, American University in the Emirates, Dubai, UAE

Swati Vijay Shinde
Pimpri Chinchwad College of Engineering, Pune, Maharashtra, India

Series Editor Fatos Xhafa
Universitat Politècnica Catalunya - BarcelonaTech, Barcelona, Spain

Academic Press is an imprint of Elsevier
125 London Wall, London EC2Y 5AS, United Kingdom
525 B Street, Suite 1650, San Diego, CA 92101, United States
50 Hampshire Street, 5th Floor, Cambridge, MA 02139, United States
The Boulevard, Langford Lane, Kidlington, Oxford OX5 1GB, United Kingdom

ISBN: 978-0-323-99412-5

For Information on all Academic Press publications
visit our website at https://www.elsevier.com/books-and-journals

Publisher: Mara Conner
Editorial Project Manager: Emily Thomson
Production Project Manager: Swapna Srinivasan
Cover Designer: Miles Hitchen

Typeset by MPS Limited, Chennai, India

Contents

List of contributors

D. Abraham Chandy
Department of Electronics and Communication Engineering, Karunya Institute of Technology and Sciences, Coimbatore, Tamil Nadu, India

Rahat Khan Afreen
DIEMS, Bamu University Aurangabad, Maharashtra, India

Auday Al-Dulaimy
School of Innovation, Design and Engineering, Mälardalen University, Västerås, Sweden

N. Anand
Department of Civil Engineering, Karunya Institute of Technology and Sciences, Coimbatore, Tamil Nadu, India

A. Diana Andrushia
Department of Electronics & Communication Engineering, Karunya Institute of Technology and Sciences, Coimbatore, Tamil Nadu, India

J. Anila Sharon
Department of Computer Science and Engineering, Karunya Institute of Technology and Sciences, Coimbatore, Tamil Nadu, India

Chandrajit Bajaj
Department of Computer Science, Center for Computational Visualization, University of Texas, Austin, TX, United States

Aparna Bannore
Department of Computer Engineering, SIES Graduate School of Technology, Mumbai, Maharashtra, India

Siew Choo Chin
Department of Civil Engineering, College of Engineering, Universiti Malaysia Pahang, Pekan, Pahang, Malaysia

Jeremy Delattre
Centre of Excellence-Interiors, Alstom, Valence, France

Jyoti B. Deone
Ramrao Adik Institute of Technology Navi Mumbai & BAMU University, Mumbai, Maharashtra, India

Sujitha Juliet Devaraj
Department of Computer Science and Engineering, Karunya Institute of Technology and Sciences, Coimbatore, Tamil Nadu, India

Amal El Arid
School of Arts and Science, Lebanese International University, Beirut, Lebanon

Mohamed Elhoseny
College of Computer Information Technology, American University in the Emirates, Dubai, UAE

Kirubakaran Ezra
Department of Computer Science and Engineering, Karunya Institute of Technology and Sciences, Coimbatore, Tamil Nadu, India

Vijay Ram Ghorpade
Bharati Vidyapeeth's College of Engineering, Kolhapur, Maharashtra, India

D. Jude Hemanth
Department of ECE, Karunya Institute of Technology, Chennai, Tamil Nadu, India

A. Hepzibah Christinal
Department of Mathematics, Karunya Institute of Technology and Sciences, Coimbatore, Tamil Nadu, India

Viraj R. Jadhao
Stevens Institute of Technology, NJ, United States

Swati Jaiswal
SCOPE, VIT University, Vellore, Tamil Nadu, India

Vijay Jeyakumar
Department of Biomedical Engineering, Sri Sivasubramaniya Nadar College of Engineering, Chennai, Tamil Nadu, India

Renu Kachhoria
Department of Computer Engineering, Pimpri Chinchwad College of Engineering, Pune, Maharashtra, India

Sanaa Kaddoura
College of Technological Innovation, Zayed University, Abu Dhabi, United Arab Emirates

R. Kanthavel
Department of CSE, King Khalid University, Kingdom of Saudi Arabia

Arijit Karati
Department of Computer Engineering, National Sun Yat-sen University, Sizihwan, Kaohsiung, Taiwan

Meghana Lokhande
Department of Computer Engineering, Pimpri Chinchwad College of Engineering, Pune, Maharashtra, India

Eva Lubloy
Department of Construction Materials and Engineering Geology, Budapest University of Technology and Economics, Budapest, Hungary

Geo Kingsly Lynus
Department of Design to Cost, Alstom Transport, Bengaluru, Karnataka, India

Gurucharan Marthi
Department of Neurology and Neurosurgery, McGill University, Montreal, QC, Canada

Mervin Ealiyas Mathews
L&T Edutech, Larsen and Toubro Limited, Chennai, Tamil Nadu, India

Rachana Y. Patil
Department of Computer Engineering, Pimri Chinchwad College of Engineering, Pune, Maharashtra, India

Yogesh Patil
IMEI Department, VBK Infrastructures, Pune, Maharashtra, India

K. Rama Abirami
Department of Information Science and Engineering, Dayananda Sagar Academy of Technology and Management, Bengaluru, Karnataka, India

Jay Rodge
NVIDIA AI, San Francisco, CA, United States

Prashanth Sali
ECE Department, Dayananda Sagar Academy of Technology & Management, Bengaluru, Karnataka, India

S.K.B. Sangeetha
Department of CSE, SRM Institute of Science and Technology, Vadapalani Campus, Chennai, Tamil Nadu, India

S. Saraswathi
Department of Computer Science and Engineering, Sri Sivasubramaniya Nadar College of Engineering, Chennai, Tamil Nadu, India

Vanlin Sathya
System Engineer, Celona Inc., Cupertino, CA, United States

Amit Sadanand Savyanavar
School of Computer Engineering & Technology, Dr. Vishwanath Karad MIT World Peace University, Pune, Maharashtra, India

R. Senthil Kumaran
Zoho Corporation Private Limited, Chennai, Tamil Nadu, India

Anandu E. Shaji
Department of Civil Engineering, Mar Baselios Christian College of Engineering and Technology, Kuttikanam, Kerala, India

Swati Vijay Shinde
Pimpri Chinchwad College of Engineering, Pune, Maharashtra, India

Shilpa Shyam
Department of Computer Science and Engineering, Karunya Institute of Technology and Sciences, Coimbatore, Tamil Nadu, India

M.N. Sumaiya
ECE Department, Dayananda Sagar Academy of Technology & Management, Bengaluru, Karnataka, India

G.R. Supreeth
ECE Department, Dayananda Sagar Academy of Technology & Management, Bengaluru, Karnataka, India

R. Supreeth
Anglia Ruskin University, Cambridge, United Kingdom

K. Veningston
Department of CSE, National Institute of Technology, Sri Nagar, Jammu and Kashmir, India

J. Vineeth
ECE Department, Dayananda Sagar Academy of Technology & Management, Bengaluru, Karnataka, India

CHAPTER 1

Introduction to different computing paradigms: cloud computing, fog computing, and edge computing

Swati Vijay Shinde[1], D. Jude Hemanth[2] and Mohamed Elhoseny[3]

[1]Pimpri Chinchwad College of Engineering, Pune, Maharashtra, India [2]Department of ECE, Karunya Institute of Technology, Chennai, Tamil Nadu, India [3]College of Computer Information Technology, American University in the Emirates, Dubai, UAE

1.1 Introduction

Data is the lifeblood of the modern age, providing significant business insight as well as real-time control over crucial business processes and activities. Businesses nowadays are flooded with data, and massive amounts of data may be routinely acquired from sensors and the Internet of Things (IoT) devices working in real time from remote locations and hostile operating environments practically from anywhere in the world. The IoT lets a device connect to the internet – a concept that has the potential to drastically alter our lives and workplaces. The IoT is predicted to grow faster than any other category of connected systems. The virtual data stream is also responsible for radically altering the way in which firms handle computing [1].

Fig. 1.1 shows the statistics about the number of IoT-connected devices worldwide in 2018, 2025, and 2030 according to the reports of Statista [2]. As per these statistics, at the end of the year 2018, there were 22 billion IoT-connected devices around the world, and by the year 2030, around 50 billion IoT devices are forecasted to be used, resulting in a massive web of interconnected devices spanning everything from smartphones to kitchen appliances.

Fully automated systems are being increasingly deployed in almost all locations to monitor and manage the associated infrastructural components, which has been possible with key technologies like the IoT and cloud computing. The cloud computing network is nothing but a set of high computing servers situated at a remote location made available for its users. IoT sensors record the different physical parameters and pass these to the cloud computing network. The cloud servers apply machine learning algorithms to the data and make powerful predictions based on which actions are taken. However, these IoT devices are constrained by limited energy as the cloud computing nodes are located far from the data source, for which delays and latencies are introduced in reporting to user queries. Much research has been carried out to overcome this limitation, with researchers proposing solutions such as edge computing and fog computing technologies. Both edge computing and fog computing make the lighter versions of cloud computing, which is easily accessible to the users with improved latencies and other features.

This chapter summarizes the computing paradigms with their architecture, advantages, drawbacks, security issues, etc. The main objective of this chapter is to set the context and provide background details that are necessary to understand the further chapters in the book.

Intelligent Edge Computing for Cyber Physical Applications. DOI: https://doi.org/10.1016/B978-0-323-99412-5.00005-8

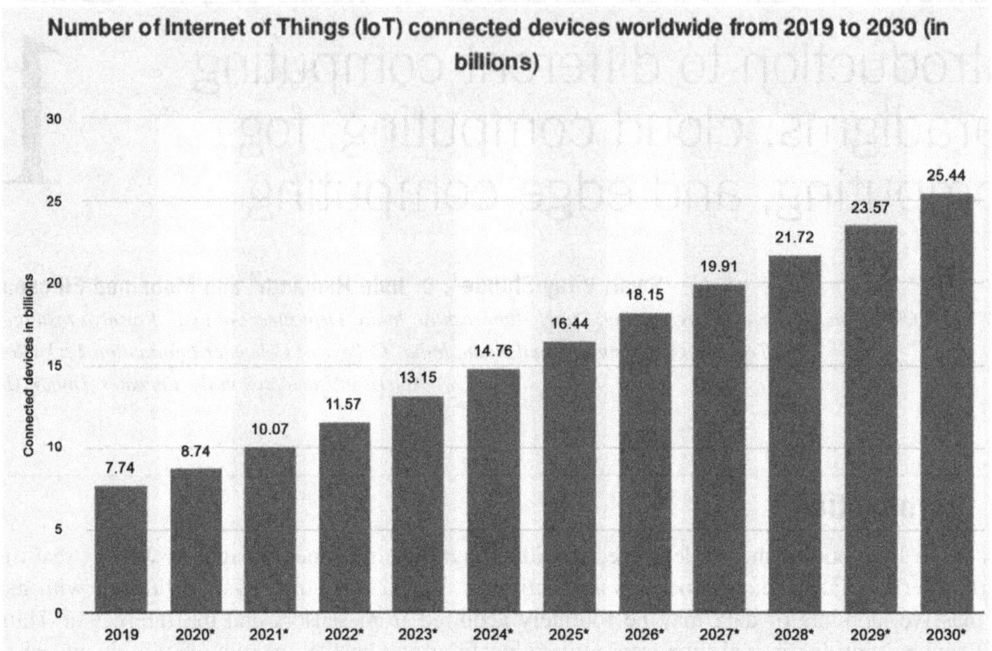

FIGURE 1.1

Rising trend in IOT devices.

(Source: https://www.mdpi.com/1424-8220/22/3/995)

The organization of the chapter is as follows: Section 1.2 summarizes the computing paradigms in brief and their comparison; Section 1.3 describes cloud computing in detail with the architecture, advantages, and drawbacks; Section 1.4 summarizes the fog computing paradigm with its pros and cons and security issues; Section 1.5 explains edge computing in detail with its background, necessity, architectures, advantages, drawbacks, and use cases; Section 1.6 highlights the challenges of edge computing implementation; and finally Section 1.7 concludes the chapter.

1.2 Computing paradigms: cloud, fog, and edge computing

Cloud computing provides the possibility to outsource the data processing and storage to a bigger and more powerful pool of servers with ease and per the requirement. This allows the users to have low resources on their devices. Cloud computing has been widely accepted in the industry for a variety of application needs as it facilitates the computing, storage, and network infrastructure capabilities getting extended with optimal cost savings.

Cloud computing enables businesses with a ubiquitous, convenient, and on-demand services platform. This has led to lesser investments by individuals and businesses, allowing them to focus on their goals rather than computing infrastructures.

Data is generated by IoT devices and saved in the cloud, which necessitates faster processing and responses. As a result, the closer the user is to the cloud, the faster the performance speed of the device [3]. Transmission delay increases as the distance between the cloud and the IoT device user grows. Because of the increased transmission delay, IoT device users may experience certain performance issues. Thus the traditional cloud computing architecture based on a centralized data center and the internet is not well suited for applications that require continuous transferring of the expanding stream of real-world data. Bandwidth constraints, latency concerns, and unpredictability in network interruptions can all work together to standoff such initiatives.

Fog computing, which refers to extending cloud computing to the network edge [4], was introduced by Cisco to address the performance issues in cloud computing. The architecture pushes intelligence down to the LAN level of network design, where it processes data in the IoT gateway or fog node. Simply put, it entails bringing the computers closer to the sensors they are communicating with [5]. However, the design of fog computing depends on multiple links in a communication chain in transporting data from our physical world assets into the digital world of information technology. Each of these links may be a potential point of failure [6,7].

The issues in fog computing have expanded the scope of edge computing that demands data processing at the network edge. Edge computing has been growing fast since 2014 with many applications and the potential to minimize latency and bandwidth charges, addressing the limitation of the computing capability of the cloud data center, boosting availability, and safeguarding data privacy and security [8].

Fig. 1.1 [9] depicts the pyramid structure of these paradigms in terms of the number of devices engaged at each level. It is clearly seen from this figure that billions of edge devices are interacting with millions of fog devices, which in turn are communicating with thousands of cloud data centers. The computational capacity of cloud computing is the highest, and that of edge computing is the lowest (Fig. 1.2).

Table 1.1 provides a comparison among the three computing paradigms.

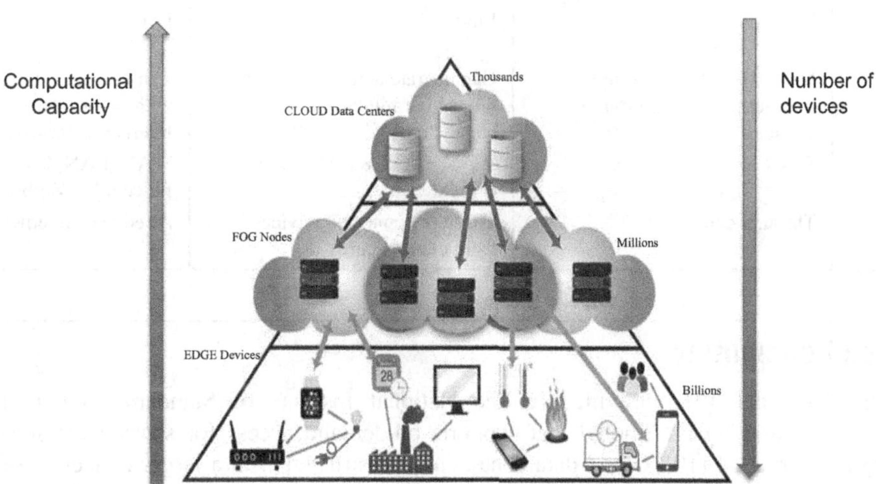

FIGURE 1.2

Recent computing paradigms.

(Source: https://www.mdpi.com/2673-4001/2/4/28)

Table 1.1 Comparison of different computing paradigms.

Attributes	Cloud computing	Fog computing	Edge computing
Operations	Cloud service providers	Users and cloud service providers	Local infrastructure providers and local businesses
Service type	Global	Less global	Local
Devices	Large-scale data centers with virtualization capacity	Devices with virtualization capacity (servers, routers, switches, access points)	Edge devices with computing capability
Computing resource availability	High	Moderate	Moderate
User proximity	Far	Relatively close	Close
Application type	Computation intensive	High computation with low latency	Low latency computation
Architecture	Centralized/hierarchical	Decentralized/hierarchical	Localized/distributed
Availability	High	High	Average
Latency	Relatively higher	Low	Low
Security	Must be provided on cloud-to-things continuum	Must be provided on participant nodes	Must be provided on edge devices
Some use cases	Scalable data storage, virtualized Application, distributed computing for large datasets	IOT-connected vehicles, smart grid/smart city, Healthcare, smart delivery using drones	Local video surveillance, video caching, traffic control
Server location	Installed in large dedicated buildings	Can be installed at the edge or in dedicated locations	Near edge devices
Power consumption	Relatively high	Low	Low
Internet connectivity	Must be connected to the Internet for the duration of services	Can operate autonomously with no or intermittent Internet connectivity	Can operate autonomously with no or intermittent Internet connectivity
Hardware connectivity	WAN	WAN, LAN, WLAN, Wi-Fi, cellular	WAN, LAN, WLAN, Wi-Fi, cellular, ZigBee
Service access	Through core	Through connected devices from edge to core	At edge of Internet

1.3 Cloud computing

According to the definition provided by the National Institute of Standards and Technology (NIST), cloud computing is a model that supports on-demand access for sharing the storage and computing infrastructure [10]. Cloud data centers are constructed with a larger collection of virtualized resources that are highly accessible and reconfigurable to the increasing workloads and support the pay-as-per-use cost model [11], making it a more convenient and preferable choice among the users. Google, Microsoft, and Amazon cloud service providers are the major giants in this domain and provide powerful cloud infrastructure for user services.

1.3.1 Architecture

The cloud architecture is as shown in Fig. 1.3, with the front end referring to client management, which acts as the graphical user interface (GUI), and the back end referring to application, service, management, runtime cloud, storage, infrastructure, and security. Each layer in the backend architecture is essential for completing the action over the cloud. The internet is the bridge to establish the connection between the front end and the back end.

- *Application*: It acts as the platform for the user to get services that are provided by the cloud
- *Services*: The cloud services are managed according to the client's requirements, which are-
  - Software as a Service (SaaS): It is a platform-independent service in which the user can go to the desired application per the requirement, for example, the Google storage system.
  - Platform as a Service (PaaS): It is a platform-dependent system that provides the platform for making an application for the client, for example, OpenShift
  - Infrastructure as a Service (IaaS): This is responsible for managing application data, middleware, runtime, for example, AWS
- *Management*: It coordinates among all the layers in the backend for communication.
- *Run-time cloud*: It provides the virtual environment for execution and computing.
- *Storage*: It provides storage infrastructure for customers.

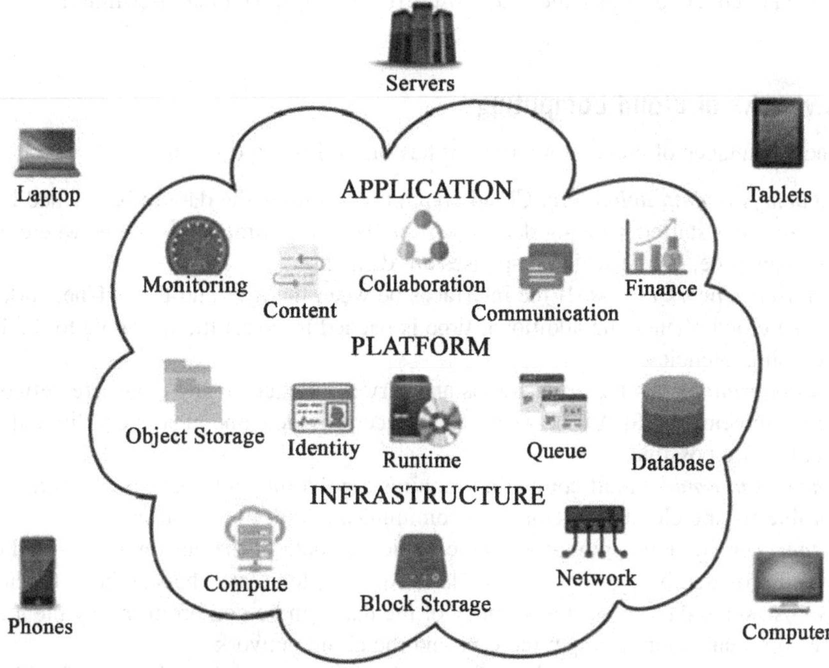

FIGURE 1.3

The architecture of cloud computing.

(Source: https://www.mdpi.com/2079-9292/10/15/1811)

- *Infrastructure*: It refers to both the software and hardware components of the cloud, including the devices for network connection and software for virtualization.
- *Security*: It provides confidentiality to clients' data to prevent attacks.

1.3.2 Advantages of cloud computing

- *Storage cost*: Cloud computing allows clients to spend less on storage as they no longer need storage disks as the cloud provides massive storage space.
- *Increased computing power*: Access to cloud computing offers access to the enormous computing power of the data center since the clients are no longer limited by the capacity of the desktop computer.
- *Collaboration*: Cloud computing allows clients located in different locations to conveniently and securely collaborate with each other. These collaborations can be internal across departments and also externally with clients.
- *Reduced software costs*: Clients do not need to purchase the software as cloud computing provides access to software per the requirements. Also, cloud computing updates the software automatically, and users do not have to worry about the software.
- *Scalability*: Clouds are easily scalable, so clients can add and remove resources per their needs.
- *Reliability and recovery*: As the cloud maintains a lot of redundant resources, failures can be effectively handled. Also, it provides the most efficient recovery plan after failures.

1.3.3 Drawbacks of cloud computing

In spite of the advantages of cloud computing, it has the following drawbacks:

- *Increased delays in data uploading*: Cloud computing requires the data to be fetched before any data processing has started, causing delays, especially in real-time applications where data uploading takes time, due to which responses are delayed.
- *Latency in user network access*: If the interfaces between the user and the IoT network are hosted on the cloud, then some additional time is needed to direct the user data to the IoT network causing latencies.
- *Limited customizations*: As the applications and services hosted on the cloud are defined by the service level agreements (SLAs) between the service providers and customers, limited customizations are possible.
- *Dependency on internet*: Cloud computing requires a good internet connection, without which it is not possible for the cloud to operate and communicate with the end-user.
- *Security concerns*: In cloud computing, the confidential data of the customer is hosted on the remote cloud provided by the service providers. As such, there are chances that malicious cloud providers misuse the data. Also, the security of the data can be compromised by the attacker during the communication between the user and the cloud network.
- *Technical support and issues*: Cloud service providers are required to give 24×7 technical support for customer queries. Many service providers are making customers rely on FAQs and online help. Also, the cloud network may experience outages or other technical issues, which need to be monitored continuously and solved immediately.

1.4 Fog computing

Cloud computing has facilitated the on-demand delivery of IT resources like storage, computing power, hardware, etc. In spite of these advantages, latency is a major problem with cloud computing. All IoT devices send data to the cloud for further processing and storage [12]. The IoT environments are constrained with respect to bandwidth, processing, memory, energy, etc., as a lot of data is sent to the cloud, and a lot of energy is invested into it [13]. So, the idea of fog computing comes into the picture, which extends the cloud nearer to the IoT devices [14].

As per the statistics, 40% of the world's data comes from sensors alone, and 90% of the world's data are generated only during the period of the last 2 years. There are almost 250 million connected vehicles worldwide and 30 billion IoT devices [15].

The ability of the current cloud model is insufficient to handle the requirement of the IoT because of the volume of data, latency, and bandwidth. Volume refers to the huge amount of data produced by different end-user applications; latency refers to the time taken by a packet for a round trip that causes the delay, which is not acceptable for time-sensitive applications; and bandwidth refers to the bit rate during transmission wherein if all the data generated by IoT devices are sent to the cloud for storage and processing then traffic will be so heavy that it will consume almost all bandwidth [16].

The primary aim of fog computing is to solve the problems faced by cloud computing during IoT data processing. Fog computing adds an intermediate layer between the cloud and edge devices, as shown in Fig. 1.4. Fog computing networks provide cloud-like services to edge devices in a distributed fashion.

1.4.1 Fog computing architecture

Fig. 1.4 depicts the fog-to-cloud layered architecture that integrates heterogeneous cloud and fog networks into a hierarchical structure. Based on the user requirements, appropriate fog is selected, and the user's request is processed. Fog computing aims to minimize the delay by using service atomization. In service atomization, services are divided into atomic subservices and executed in parallel.

The objective of the service allocation process is to reduce the allocation delay, load balancing, and energy-usage balance among the distinct fogs. Fog nodes in the fog layer communicate with each other to allow the IoT to utilize its maximum potential and reduce the resulting delays.

Fog computing is a distributed paradigm that provides cloud-like services to the network edge [6]. Studies on fog computing are still premature, leading to many emergent architectures for computing, storage, control, and networking, which distribute these services closer to end-users.

The bottom layer in Fig. 1.4 includes all IoT devices like sensors, mobile phones, smart home devices, smart traffic signaling systems, smart cars, factories, aerial devices, etc. These devices sense the surrounding environment and read the data. This data is passed on to the fog layer.

The fog layer, present between the edge devices and the cloud network, has devices like routers, gateways, access points, base stations, fog nodes, etc. Fog nodes can be static, such as those located in some office premises, or dynamic, such as those fitted in moving vehicles. Fog nodes can temporarily store, compute, and transfer client data and services. They are connected to the cloud data centers using Internet Protocol and forward the requests to the cloud network for enhanced processing and storage.

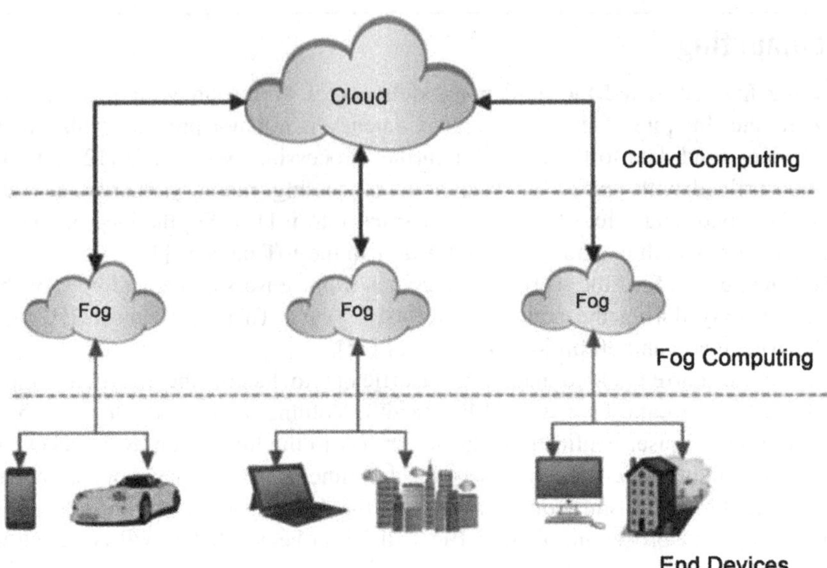

FIGURE 1.4

The architecture of fog computing.

(Source: https://www.google.co.in/url?sa=i&url=https%3A%2F%2Fwww.mdpi.com%2F2504-2289%2F2%2F2%2F10&psig=AOvVaw0sDOCq5Mcfhg4hPVFbK_u3&ust=1671775879961000&source=images&cd=vfe&ved=0CBAQjRxqFwoTCPjK39DIjPwCFQAAAAAdAAAAABAI)

The Cloud layer provides data centers for permanent storage of the data and high-speed speed computing services to the clients. The data centers are scalable and provide on-demand services to their users. The data that is not needed in the proximity of clients is stored in the cloud data centers.

1.4.2 Security issues in fog computing

Some additional security concerns are introduced in fog computing due to geo-distribution, mobility, and heterogeneity. The following subsections describe the different attacks and privacy issues that need to be considered in the context of fog computing.

1.4.2.1 Attacks in fog computing

The following attacks are possible in fog computing networks [13]:

• Forgery	• Eavesdropping	• Impersonation
• Tampering	• Denial of Service (DoS)	• Virtual machine attack
• Spam	• Man in the Middle	• Side-channel attack
• Sybil	• Collusion	• Session hijacking
• Jamming		

1.4.2.2 Privacy issues in fog computing

Each user wants their data to be safe and secure over the wireless medium, but, unfortunately, it is very difficult to preserve the privacy of the user's confidential data in wireless networks. Privacy preservation is an important issue in fog computing because the confidential data of users is exchanged, gathered, processed, and transmitted over fog nodes.

Following are the six important aspects of privacy in fog computing [13]:

1. User privacy
2. Identity privacy
3. Data privacy
4. Usage privacy
5. Location privacy
6. Network privacy

1.5 Edge computing

The explosion of IoT devices at the network edge generates vast amounts of data that must be calculated at data centers, pushing network bandwidth requirements to their maximum. With network technology advancements, data centers cannot ensure appropriate transfer rates and response times, which may be a significant requirement for certain applications.

The goal of edge computing is to transfer processing away from data centers and toward the network edge, where devices, mobile phones, or network gateways may conduct processing and deliver services on behalf of the cloud. An edge device is a piece of hardware that regulates data flow at the intersection of two networks [17]. Cloud computing and the IoT have boosted the role of edge devices, necessitating greater intelligence, computing, and artificial intelligence (AI) services at the edge network. These devices can be as small as smart thermostats, smart doorbells, home cameras or cameras on automobiles, and augmented reality or virtual reality glasses or can also be as large as industrial robots, automobiles, smart buildings, and oil platforms.

Both fog and edge computing seem similar as they both intend to bring computing near the end-user. However, fog computing places computing at the local area network, while edge computing places the computing part on end-user devices themselves. In fog computing, edge devices send data to the LAN gateways where it is processed, and the response is sent back to edge devices. But in the case of edge computing, computations are performed near the edge of the network, which is closer to the data source.

1.5.1 Architecture of edge computing

The management of IoT networks is complex due to the heterogeneity of its resources, which causes problems with communication protocols, real-time operations, data management, large data storage, security, and privacy [16]. Edge computing architectures provide a solution to IoT infrastructures in this area since they can manage the diverse data generated by IoT devices. Fig. 1.5 depicts a simple three-layered edge computing design.

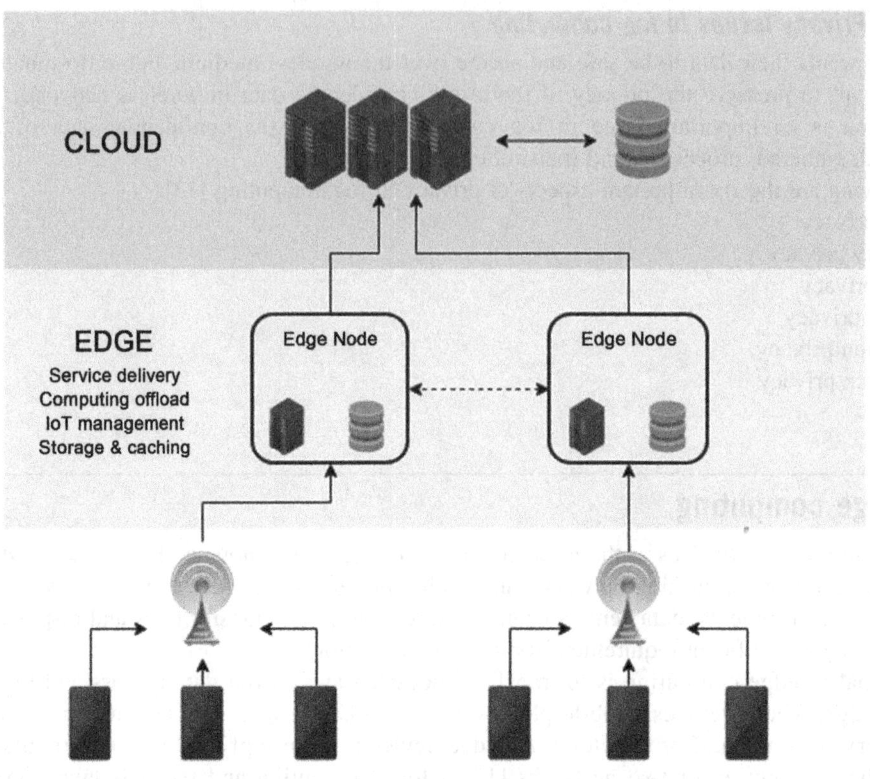

FIGURE 1.5

Three-layered architecture of edge computing.

(Source: https://www.wikiwand.com/en/Edge_computing)

Layer I: The first layer is in charge of data ingestion and the operations involved and contains both IoT equipment (sensors, smart meters, smart plugs, etc.) and users.

Layer II: Edge nodes, comprising the second layer, are in charge of data processing, routing, and computing.*Layer III*: This layer is made up of several cloud services that have higher computing requirements and is in charge of tasks such as data analytics, artificial intelligence, machine learning, and visualization, among others.

1.5.1.1 Applications

Edge devices in layer-I generate enormous data, which is moved across LAN/WAN to the edge nodes/servers. The edge nodes/servers are general-purpose computers located in client premises, such as factories, hotels, hospitals, companies, banks, etc. Generally, the computing capacity of edge servers in terms of the number of cores is 8, 16, or more, and a memory of 16 GB. Typically edge servers are used to run enterprise applications. Edge nodes perform the functions of data processing and reduction, data caching, control responses, and virtualization.

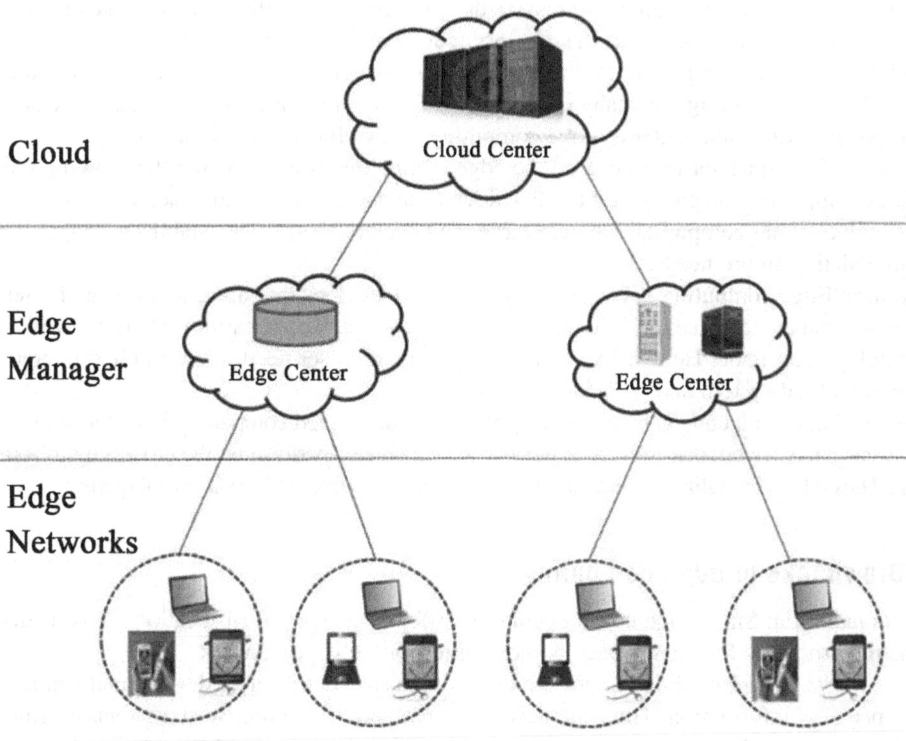

FIGURE 1.6

Edge infrastructure.

(Source: https://www.researchgate.net/figure/Architecture-of-edge-computing_fig2_333230540)

Edge nodes provide the temporary caching of the data and a small amount of immediate computing. This is extremely useful in real-time applications as the data and processing requests need not have to travel to cloud networks. Bigger data processing requests and storage demands are forwarded to the cloud data centers by the edge gateways. Cloud network provides the heavy computing and permanent storage of the data.

Edge infrastructure is located near the end-user network and brings storage and processing closer to the data source as given in Fig. 1.6. In comparison to the cloud, edge infrastructure is located in smaller facilities close to the end-user network. By bringing compute, storage, and network resources closer to the source of the data, edge data centers offer various benefits to both organizations and their workloads [17].

1.5.2 Advantages of edge computing

- *Speed*: Edge computing reduces latency by facilitating IoT edge computing devices to process data locally or in nearby edge data centers without accessing the data centers of traditional cloud computing.

- *Security*: Centralized cloud computing systems are vulnerable to distributed denials of services or attacks and power outages. On the contrary, edge computing systems are more secure as they distribute the processing among multiple devices and data centers. However, the distributed nature of edge computing has many entry points of malware, but they can be easily sealed by the respective edge data center as edge computing allows functioning individually.
- *Scalability*: It is much easier to extend the edge computing systems as per the growing business needs as companies can easily add the IoT devices and edge data centers according to the future needs, unlike cloud computing, wherein large data centers need to be established for accommodating future needs.
- *Versatility*: Edge computing is based on edge and IoT devices that are always on and interacting with edge data centers making it a versatile system for many applications where network reachability is an issue. Here unlike cloud computing, the user need not log in and interact with the central cloud system and wait for a response.
- *Reliability*: Edge computing is based on the principle of distributed computing. Even though some edge data centers have failed, still, edge devices are capable of performing the processing to some extent. Thus edge computing systems are more reliable compared with their counterparts.

1.5.3 Drawbacks of edge computing

- *Hard to maintain*: Since edge nodes operate in different geographical locations, it is difficult to define the processes for monitoring and controlling of the edge network.
- *Increased attack surface*: Edge nodes are directly interfaced with user devices and can access users' personal information. The edge network is dynamic in nature, so fixed security rules cannot be defined. Also, many edge nodes are operating, and if any of the edge nodes is compromised, then security will be at stake.
- *Not ideal for heavy workloads*: Heavy workloads can impact the performance of edge networks. It is better to process bigger tasks needing voluminous data processing on the public cloud data center.
- *Cost*: Setting up an edge network is costly as it requires additional and advanced infrastructure.

1.5.4 Security in edge computing

In edge computing the following security challenges need to be considered [18]:

- Edge nodes are very much nearer to the end-users and thus receive a large amount of users' personal data. So, there are more chances of compromising this data.
- Limited resources are available with edge computing than cloud computing, thus limiting the use of heavy algorithms for encryption and other security algorithms.
- Edge computing network is dynamic in nature, making it difficult to set up the security rules for the network. So, attackers can easily join such dynamic networks and compromise security.
- Edge computing possesses limited network resources compared with cloud computing; as such, they do not support complex encryption algorithms.
- Different factors lead to security and privacy issues in edge computing networks that put users' personal data at risk, as discussed below:

1.5.4.1 Attacks on edge computing

The following security attacks at different levels of infrastructure are possible in edge networks:

Core Side Edge Servers	• Privacy Leakage • Data Tampering • DoS • Service Manipulation
Edge Servers	• Privacy Leakage • DoS • Privilege Escalation • Service Manipulation • Rogue Data Center • Physical Damage
Edge Networks	• DoS • Man-in-the-middle • Rogue Gateway
Mobile Edge Devices	• Data Injection • Service Manipulation

1.5.4.2 Privacy issues in edge computing

The following privacy issues need to be considered in edge computing networks:

- Strengthening the security measures to protect the vulnerability of the network from malicious users and edge nodes
- Susceptible communication among edge devices
- Data recovery after a system failure
- Mobility of edge devices
- Lack of predefined network visibility and user data selection

1.5.5 Use cases of edge computing

The increasing amount of IoT devices has resulted in a number of applications where edge computing is implemented. Fig. 1.7 depicts some of the applications of edge computing [16].

A few of the use cases of edge computing are discussed below [17]:

1. *Autonomous vehicles*: While autonomous vehicles are not yet ready for the masses, their feasibility would be several years away without edge computing techniques. Vehicles can exchange real-time sensory data and corroborate and enhance choices with fewer onboard resources, minimizing the rising cost of autonomous AI systems.
2. *Healthcare*: In a hospital, edge nodes can be used to analyze data from multiple monitoring devices to ensure that analysis is completed in a timely manner.
3. *Telecommunications*: A communications service provider can give real-time statistics for video streaming via edge nodes such as a 5 G router.
4. *Point of sale system*: Edge nodes aggregate and process data from sales terminals, lowering the cost of transmitting raw data to the cloud for processing.

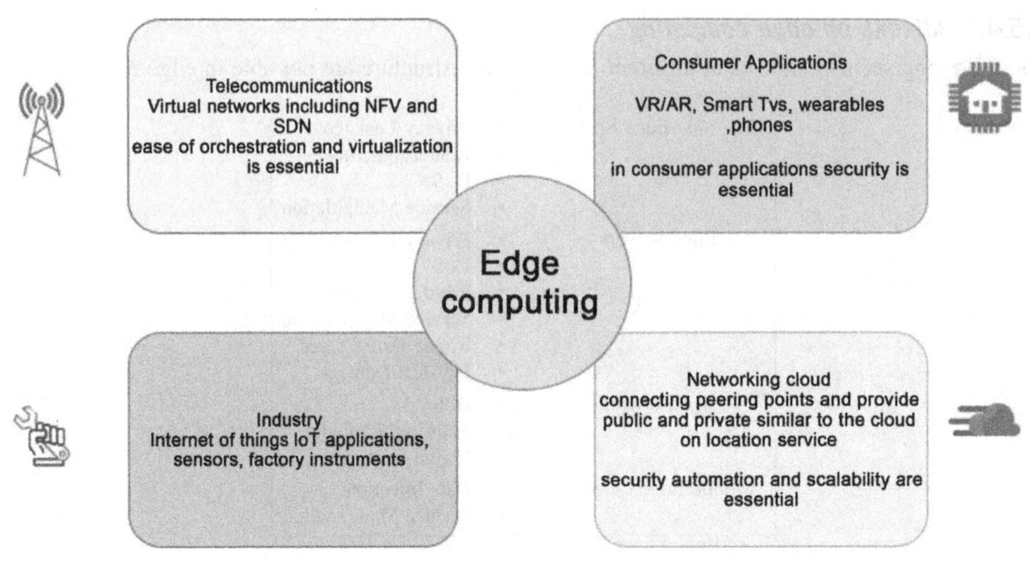

FIGURE 1.7

Applications of edge computing.

(Source: https://www.researchgate.net/publication/356924481_An_Overview_of_Fog_Computing_and_Edge_Computing_Security_and_Privacy_Issues/figures)

5. *Supply chain tracking*: Edge nodes let distribution centers collect and analyze high-bandwidth data to process orders more efficiently.
6. *Remote monitoring of oil and gas plants*: Generally oil and gas plants are located in remote places, and their leakages can cause major accidents and losses. So, they need to be monitored and processed in real time. Edge computing enables faster processing and good monitoring control over their activities.
7. *Transport and traffic monitoring*: Edge computing enables optimized utilization of public transport through route selection, managing extra lanes, and deciding bus and train frequencies. Real-time traffic data can be transferred to edge devices, and traffic control can be obtained without transferring this data to the centralized cloud.
8. *Smart grids for effective energy consumption*: Industries can set up smart grids to monitor energy usage and plan effective usage.
9. *Predictive maintenance*: With the help of edge computing, manufacturers can monitor the changes in the machine functionality before failure and mitigate them before getting more serious. This is possible because edge computing has brought computing closer to the machines, and sensors can track the changes and pass them to edge nodes.
10. *Smart homes*: Edge computing has enabled the processing and storage of home data acquired through home sensors around the home only, reducing latency and helping to get faster responses for user commands. For example, the Alexa system answers quickly as the data need not have to travel to the cloud servers.

1.6 Challenges of edge computing

Fog computing and edge computing are relatively new technologies. Tested frameworks for implementing these computing systems are under research, which are required to work in real-time scenarios. Current cloud computing frameworks support high computing tasks, but they need to be finetuned for real-time applications.

- *Security in terms of authentication of edge devices* and also of cloud service providers needs to be provided as billions of edge devices are connected to the network and are continuously increasing day by day.
- *Privacy* is to be ensured as these computing paradigms are dominated by wireless connectivity, in which chances of data leakage during communication exist.
- *Interconnection of computing nodes* and heterogeneous edge devices to ensure secure communication among them needs to be taken care of.
- *Placement of edge and fog servers* to deliver maximum service to the clients is a challenging task.
- *Energy consumption* should be reduced as the tasks are distributed among the multiple fog nodes and edge nodes.
- *Performance management* in real-time scenarios is to be ensured.
- *Discovery of edge nodes* needs to be effectively implemented to allocate clients' tasks to the appropriate edge nodes.

1.7 Conclusion

This chapter has summarized the recent computing paradigms, including cloud computing, fog computing, and edge computing. There is a tremendous rise in the number of users of the internet and its services. Cloud computing is evolving continuously to satisfy the huge demand of the users and provide seamless and faster service to the end-users. Cloud computing has opened the door for many businesses to function with bare-minimum investments. However, connecting each user request to the centralized cloud network generates a lot of traffic and decreases the quality of service. Fog computing solves this problem by filtering the data and allowing only the necessary accesses to the cloud. Edge computing has offered a finer version to this problem by providing computing in the user premises itself, which has become possible by segregating the tasks to be executed by the edge nodes and the task to be sent to fog nodes or to the cloud data centers. The proper allocation of services to edge, fog, and cloud servers ensures maximum throughput and minimum latency, and reduced energy consumption. However, there are some challenges that need to be taken care of, which have opened up many research opportunities as edge and fog computing are still in their infancy.

References

[1] M. Henze, R. Matzutt, J. Hiller, M. Erik, J.H. Ziegeldorf, J. van der Giet, et al., Complying with data handling requirements in cloud storage systems. IEEE Trans. Cloud Comput. 2020, 1.

[2] https://www.statista.com/statistics/802690/worldwide-connected-devices-by-access-technology/.

[3] I.U. Din, M. Guizani, B.S. Kim, S. Hassan, M.K. Khan, Trust management techniques for the Internet of Things: a survey, IEEE Access. 7 (2018) 29763–29787.

[4] J. Kaur, A. Agrawal, R.A. Khan, Security issues in fog environment: a systematic literature review, Int. J. Wirel. Inf. Netw 27 (2020) 467–483.

[5] M. Chiang, T. Zhang, Fog and IoT: an overview of research opportunities, IEEE Internet Things J 3 (2016) 854–864.

[6] R. Priyadarshini, Kumar Barik, R. Dubey, H. Fog-SDN, A light mitigation scheme for DDoS attack in fog computing framework, Int. J. Commun. Syst 33 (2020) e4389.

[7] A. Aljumah, T.A. Ahanger, Fog computing and security issues: a review. In Proceedings of the 2018 7th International Conference on Computers Communications and Control (ICCCC), Oradea, Romania, 8–12 May 2018, pp. 237–239.

[8] Jagdeep Singh, Parminder Singh, S.S. Gill, Fog computing: a taxonomy, systematic review, current trends and research challenges, J. Parallel Distrib. Comput 157 (2021) 56–85. Available from: https://doi.org/10.1016/j.jpdc.2021.06.005. ISSN 0743-7315.

[9] R.S. Dikhit, All about edge computing architecture, open source frameworks and IoT solutions, March 2018. https://www.opensourceforu.com/2018/03/all-about-edge-computing-architecture-open-source-frameworks-and-iot-solutions/.

[10] J. Huttunen, J. Jauhiainen, L. Lehti, A. Nylund, M. Martikainen, O. Lehner, Big data, cloud computing and data science applications in finance and accounting, ACRN Oxf. J. Financ. Risk Perspect 8 (2019) 16–30.

[11] M. Heck, J. Edinger, D. Schaefer, C. Becker, IoT applications in fog and edge computing: where are we and where are we going? In Proceedings of the 2018 27th International Conference on Computer Communication and Networks (ICCCN), Hangzhou, China, 30 July–2 August 2018; pp. 1–6.

[12] S.-Y. Lin, Y. Du, P.-C. Ko, T.-J. Wu, P.-T. Ho, V. Sivakumar, Rama subbareddy, Fog Computing Based Hybrid Deep Learning Framework in effective inspection system for smart manufacturing, Computer Commun. 160 (2020) 636–642. Available from: https://doi.org/10.1016/j.comcom.2020.05.044. ISSN 0140-3664.

[13] A.M. Alwakeel, An overview of fog computing and edge computing security and privacy issue, Sensors 21 (2021) 8226. Available from: https://doi.org/10.3390/s21248226.

[14] K. Sha, T.A. Yang, W. Wei, S. Davari, A survey of edge computing-based designs for IoT security, Digit. Commun. Netw 6 (2020) 195–202.

[15] A. Alwarafy, K.A. Al-Thelaya, M. Abdallah, J. Schneider, M. Hamdi, A survey on security and privacy issues in edge computing-assisted internet of things, IEEE Internet Things J 8 (2020) 4004–4022.

[16] Y. Mao, C. You, J. Zhang, K. Huang, K.B. Letaief, A survey on mobile edge computing: the communication perspective, IEEE Commun. Surv. Tutor 19 (2017) 2322–2358 [CrossRef].

[17] W. Masri, I.A. Ridhawi, N. Mostafa, P. Pourghomi, Minimizing delay in IoT systems through collaborative fog-to-fog (F2F) communication, Ninth Int. Conf. Ubiquitous Future Netw. (ICUFN) (2017) 1005–1010. Available from: https://doi.org/10.1109/ICUFN.2017.7993950.

[18] B. Varghese, N. Wang, S. Barbhuiya, P. Kilpatrick, D.S. Nikolopoulos, Challenges and opportunities in edge computing, IEEE Int. Conf. Smart Cloud (SmartCloud) (2016) 20–26. Available from: https://doi.org/10.1109/SmartCloud.2016.18.

CHAPTER

Supervised machine learning techniques to protect IoT healthcare environment against cyberattacks

2

Sanaa Kaddoura[1], Amal El Arid[2] and Auday Al-Dulaimy[3]

[1]*College of Technological Innovation, Zayed University, Abu Dhabi, United Arab Emirates* [2]*School of Arts and Science, Lebanese International University, Beirut, Lebanon* [3]*School of Innovation, Design and Engineering, Mälardalen University, Västerås, Sweden*

2.1 Introduction

The Internet of Things (IoT) has become a trend and a common interest of evolution in many industries. Connecting things to the Internet is the primary focus for IT specialists in next-generation technologies. It simply tells that it is time to connect humans, animals, places, and things in the world through the Internet [1]. One of the most dynamic industries, the healthcare domain, is gaining advantage from using IoT technology in their workflow, which include monitoring patients, tracking people and things in motion inside hospitals, and improving medical services and workflow in hospitals. It also includes patients' identification and authentication, automatic data collection, and stock medical management [2,3].

The healthcare domain is afflicted with the aging population with prolonged illnesses, leading to increasing demand for hospital resources, including beds, physicians, and caretakers such as nurses. In addition, almost the whole world has suffered from the coronavirus disease (COVID-19) [4]. Therefore, integrating IoT systems into healthcare systems is becoming essential to decrease the pressure on healthcare systems. Fig. 2.1 depicts the connection between the hospital, doctors, and patients using an efficient, reliable, and intelligent healthcare system.

The results are then sent to the physicians for treatment actions.

Although IoT enhances the efficiency of using services to simplify people's life and create values in various domains, in particularly in the healthcare field, this technology can reduce people's security and privacy. Cybersecurity instances are increasing threats to healthcare organizations, particularly to hospitals. Thus, it has been a big concern of the healthcare domain to provide better-secured medical services to their patients inside and outside the hospitals. Hospitals invest considerable money in protecting their IT systems [5]. Consequently, several researchers have focused on making IoT the best stress-release out of the healthcare systems based on security and data exchange efficiency [6]. According to [7], cybersecurity includes several actions: securing the network, improving the IT management of storing and accessing data, and training healthcare workers (doctors, nurses, patients, etc.).

Intelligent Edge Computing for Cyber Physical Applications. DOI: https://doi.org/10.1016/B978-0-323-99412-5.00001-0

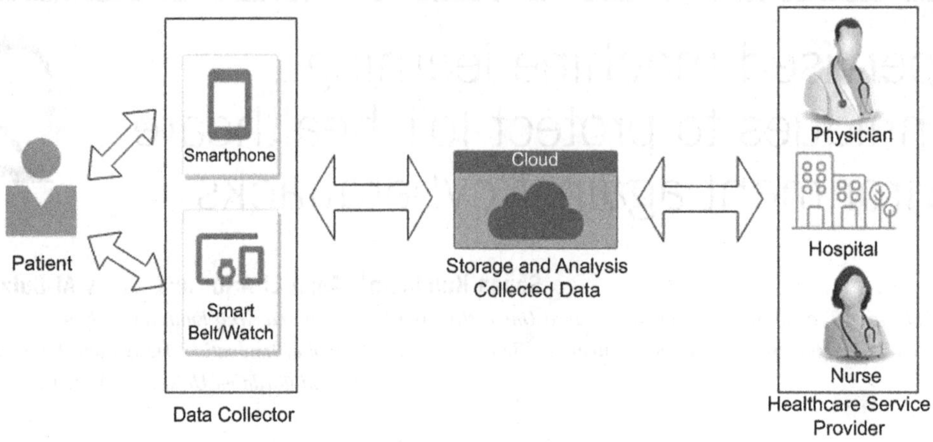

FIGURE 2.1

The Internet of Things system in the healthcare domain; data is collected from patients through smartphones and wearable smart devices, stored and analyzed in the cloud.

Coronado and Wong [8] discussed the cybersecurity issues created due to the advancement of technology in the healthcare systems. According to them, monitoring patients and collecting data became more accessible and vulnerable to threats such as malware and viruses. Moreover, it is essential to increase hospital network security, and the latter is managed. Tao et al. [9] focused on studying security concerns for monitoring patients through sensors using IoT systems because attacks can also happen in the healthcare domain; attacks can include data collision, breaching, and integrity.

Consequently, providing high security for patients' information is promising in the coming years. It enhances the healthcare systems by integrating IoT systems into their existing system [9]. Moreover, Li et al. [10] proposed a newly secured framework for edge computing in IoT-enabled healthcare systems. The IoT devices are authenticated to get permission for access. Then, the devices collect data from the patients and send them to the cloud for storage, process, and analysis. Developing a software platform ensures trust enhancing and secures systems and applications for private hospitals in Europe [5]. The authors proposed cybersecurity and privacy risk solutions to provide secured tools for data exchange. In addition, their platform aimed to increase security and privacy awareness of the employees in the healthcare sector to improve the capability and trust in exchanging and handling sensitive data securely.

The main objective of this chapter is to evaluate cybersecurity issues after the integration of IoT in health informatics and medical services. The next section represents background information about IoT in terms of definition and applications, particularly in health informatics and medical services. Moreover, this section describes cybersecurity issues in healthcare systems. Security problems in healthcare systems for the past years are discussed later, particularly during the COVID-19 pandemic. Recent research in the IoT and cybersecurity for health informatics are also represented. Supervised machine learning techniques are applied to health informatics datasets to predict malicious behavior. Some future challenges and studies related to the contribution of cybersecurity in the health informatics and medical services fields are proposed. Finally, the chapter concludes the information discussed and presented in this chapter.

2.2 Background

This section discusses background information on IoT in terms of definition and applications, in particular the Internet of Medical things (IoMT). Moreover, cybersecurity issues in healthcare systems in the past years are also revealed. In addition, some challenges in the healthcare field are tackled, showing the pros and cons of securing data in the healthcare industry.

2.2.1 Internet of things overview

IoT is a system that connects objects such as people, animals, machines, and intelligent devices. When people are connected to the Internet, they can send/receive information to/from any place in the world. In IoT, all the things that are connected can:

- Collect information from one or different parties then send it again.
- Receive information from one or different parties then act accordingly.

IoT can create information about collected data from many connected things, analyze it, and make decisions based on machine learning algorithms. Some examples of things are security cameras, vehicles, buildings, sensors, and software. These things can exchange data among each other as long as they are connected. IoT has many benefits, including:

- Minimizing human effort because the devices can communicate with each other without the intervention of people
- Saving time because intelligent things are faster in computing than people
- Improving security and efficiency among the interconnected devices, etc.

On the other hand, IoT has many drawbacks, including:

- Security because a network that consists of connected things can be attacked at any time
- Privacy because the system contains all the details about personal data
- Complexity because designing, implementing, testing, maintaining, and installing the IoT system could be very complicated.

Each thing is given a unique identification to exchange data over a network without the requirement of direct interaction from humans or computers. Because it is easy now to connect things such as cars, home appliances, baby monitors, to the Internet for data exchange, an IoT system consists of embedded systems such as real-time sensor networks, dynamic networks of embedded computing devices, Internet-enabled physical devices, and other technologies [11]. IoT systems become an interest if it consumes little amount of power because it is a key factor for the IoT system's owner. Consuming a little amount of power is achieved through careful hardware design, software design, and application algorithms. Security and safety are other key factors for the owner as well. Because privacy is a common concern for any organization or individual, securing a computer system is required at the design level.

Moreover, the physical system design should be safe from attacks. Many organizations in different sectors are adopting IoT technology for many reasons. It simplifies specialists' jobs, improves jobs' efficiency, and controls the processes of each task to be accomplished. The applications of IoT technology include wearable, traffic monitoring, fleet management, agriculture, hospitality, energy saving, water supply, and health. The latter is known as the Internet of Medical Things.

2.2.2 Internet of medical things

Most healthcare systems use manual management systems to make an appointment, prescribe medicine, follow patients' case history, etc. To prevent missing patients' demographic data, accelerate treatment processes, follow medication history records, generate billings, manage drug and medical equipment stock, the Internet of medical things (IoMT) rises up to enhance the healthcare domain and the medical services such that the field follows the growth of technologies. The IoMT platform is a combination of medical devices and applications that, using networking technologies, can connect doctors, nurses, and patients in the healthcare technology systems for information exchange [12]. It is also comprised of sensors and electronic circuits that are connected to patients' physical bodies or wearable devices, giving doctors the ability to monitor their patients outside the hospital in real-time [13]. Accordingly, doctors can take necessary actions if their patients are at high risk. Thus, the IoMT system can help many patients from fatal death or damage, if required.

Monitoring patients with neurodegenerative diseases is now applicable using a camera-based system with high definition [14]. A wireless sensor tag-based wristwatch is another intelligent system used to monitor patients with brain and neurological diseases [15]. Ali et al. [16] developed another smart system to monitor diabetic patients. The system consists of a sensor node connected to a smartphone via Bluetooth. An application on the smartphone is responsible for monitoring the glucose measurements in the patient's body. More research work was intended to focus on developing systems that remotely monitor older people who may lose conscience at any time [17].

Moreover, using the IoMT system, doctors can also monitor their patients inside the hospital. The embedding of smart beds with special sensors to perceive blood pressure, oximeter, body temperature, and vital signs can help doctors and nurses monitor their patients quickly [18]. Medical equipment can also be tracked by implanting RFID chips. Drug and medical stock can also be managed using IoMT systems to maintain medical services at their high level.

Not only patients' monitoring is necessary, but also keeping their confidential data secured is the aim. Because attacking the healthcare system may result in the healthcare organization losing its reputation, it is mandatory to make a high level of security of the organization databases. For this purpose, many research works focused on providing data transactions in a secured environment. Accordingly, researchers suggested many approaches to detect malicious data exchange. For instance, Kaddoura et al. [19] proposed an efficient damage assessment and recovery method to clean the database from infected data. Based on data dependency, their algorithm, using the matrix data structure, identifies the affected part of the database for later recovery.

2.2.3 Cyberattacks in healthcare systems

After integrating IoT technology in most industrial fields, security breaches increased due to many factors. The users' lack of knowledge in using technology and network is the primary reason for vulnerability to attacks [5,20]; the healthcare industry is not far from cyberattacks.

Cyberattacks can uncover most patient information, leading to high financial costs to recover control of patient data and overall hospital systems. For instance, more than 780 thousand patient records were stolen from the department of technology server in the State of Utah Department of Health. Around 32 thousand patient records from Saint Joseph's Health system were publicly available through Internet search engines [21]. Thus, the authors provided a primitive solution by creating patients' identities in the electronic health records to protect patients' sensitive information without impeding the workflow of healthcare clinics, particularly doctor-patient relationships. Later, many research works presented security policies, known as "norms" [22]. They proposed a framework to determine breaches based on norms, including authorization, commitment, and prohibition. According to their study, a violation occurs when violating standards used to describe the users' expected behavior.

Healthcare organizations are aware of the vulnerability of cyberattacks that could lead to losing or modifying patients' records. Moreover, many impacts are raised due to attacking patients' private information and network communication information of healthcare organizations [23]. Patients' and vulnerable network information can affect confidentiality, integrity, and availability. For instance, patients could feel upset and lose confidentiality toward the healthcare organization. The health service provider could lose reputation and dramatically impact its business. Serving patients could be substantially affected. In addition, the IoT infrastructure of smart devices used in the healthcare industry, such as control systems, sensor devices, and wearable devices, had a bad influence due to their connectivity in the healthcare data networks, leading to an increase in cyberattacks Abouzakhar [24].

Because interaction with healthcare systems is essential, patients need to access the healthcare cloud through the Internet. Thus, the network could be more vulnerable if users are not technologically educated. Cloud-siders attacks include denial of service, unauthorized access, ARP poisoning (redirecting packets for sniffing), virtual machine backdoors (using covert communication between two parties), hypervisor attacks (spreading malicious messages by obtaining high administration-level rights), rootkit attacks (creating a fake cover channel for sending malicious codes), and virtual machine escape [25–27].

2.3 Problems in healthcare systems

According to Reader et al. [28], healthcare systems receive many complaints from patients. Some of these complaints include problems in healthcare delivery; others involve the patients' safety. Consequently, patients are also addressing the healthcare systems to provide additional information on enhancing the healthcare domain. Based on the collected dataset of complaints, healthcare organizations should analyze the given data and strengthen their capabilities to provide better medical services. However, the healthcare industry faces new challenges due to rapid health informatics change and medical services, particularly during the COVID-19 pandemic. Sullivan [29] listed the six challenges that the healthcare field is facing as follows:

- Cybersecurity. This industry is still suffering from data breaches, ransomware, and other cybersecurity issues. It is always necessary to keep the patients' records secure. However, telehealth has become the first contributor to cyberattacks increase in the last years. Attacking this medical service can lead to data breaching.

- Telehealth. After the COVID-19 pandemic, many physicians have moved to telehealth medical services to minimize in-person interaction. The adoption of telehealth has increased from 11% in 2019 to 46% in 2020. However, telehealth is not an appropriate medical service for patients with Alzheimer's, for instance.
- Patient Experience. Any organization's reputation is affected by the patients' experience. Patients are always seeking better services from their providers because they are responsible for a high percentage of the service bill payment. Accordingly, keeping patients satisfied is not only by providing good service but also by giving them access to all the information they need about their health.
- Big Data. Data generated by the healthcare industry is becoming huge. However, any change made at the patient's level, for instance, changing in insurance plan, is not informed to the health organization. Accordingly, the payment method can be wrongly processed, making a disturbance for the patient, health organization, and insurance company simultaneously. Moreover, no single source is providing the healthcare system with data. It is a big mess and a waste of time for wrongly accessed data from any source.
- Invoicing and Payment Processing. Although insurance companies are sometimes responsible for payment procedures, patients are involved in paying their bills. Accordingly, it is essential to ensure that the generated billing is patient-friendly. Moreover, it is necessary to offer payment statements in a form that is much familiar to the patients (checks, credit cards, tap-to-pay, etc.).
- Price Transparency. Many patients cannot pay their medical financial responsibility because medical services' payments are not clear from the health organization part. Consequently, most health organizations should make their services price accessible to the patients and consumers so they know whether they can afford medical services in any organization. It will also help in making the payment bill much readable.

In the next section, the contribution of recent research to the above-listed problems is detailed. It is necessary to mention that improving data transactions, privacy, and security is a day-to-day activity because information technologies are frequently and rapidly evolving, particularly security issues during the COVID-19 pandemic.

2.4 Recent research in cybersecurity and health informatics

Health informatics is integrated into the healthcare industry to organize and analyze electronic health records (EHR) to enhance healthcare productivity, ensure patients' safety, and increase the quality of patient care. Alamri [30] investigated the use of IoT in EHR to help physicians accessing patients' data efficiently. The author proposed a semantic middleware system to support the semantic incorporation and functional cooperation between EHR systems and IoT health informatics systems.

Based on patients' experience in the healthcare systems, their agreement for contributing the IoT technology into the healthcare domain is necessary. A technology assessment acceptance

(TAM) collected the opinions of patients of different ages and from different genders. The survey showed that most of the patients accepted the idea of the incorporation between the IoT technology and the healthcare systems [31]. Moreover, a privacy impact assessment (PIA) was undertaken by Pribadi and Suryanegara [32]. They proposed some tools to facilitate the integration of IoT technology into healthcare systems. Based on a questionary survey in Indonesia, the patients and customers of healthcare organizations have stated their anxieties about the device, communication, visualization, and application security. Accordingly, many suggested regulations to solve security-related problems were studied, showing their cost and benefit based on the security level required.

Consequently, Saheb and Izadi [33] discussed the IoT big data analytics (IoTBDA) paradigm and its impact on the healthcare industry. Based on many research papers, the authors concluded that IoTBDA influenced information storage, retrieval, and acquisition in the healthcare domain. In their book chapter, the authors discussed three main drivers: (1) computing: a response to reduce data congestion [34], (2) convergence of storing IoT big data based on critical data (sent to fog systems), and noncritical data (sent to centralized cloud systems), and (3) convergence of data abstraction.

Paul et al. [35] implemented a context-sensitive fog computing environment was implemented to improve healthcare systems. Their system consists of the fog layer, cloud, user devices, and sensors at the user level. The computational process for data to travel across the different layers is reduced based on a distributed data transfer to multiple nodes in the network, keeping an improved security level in the network.

Smart-health systems face critical issues such as data security, transfer costs, memory storage, and privacy between different network parties. Thus, Tahir et al. [36] used blockchain technology to enhance security and improve data sharing using decentralization servers to reduce cost overhead in the system. For this purpose, they proposed a blockchain-enabled IoT network to present an authentication and authorization framework using a probabilistic model. Moreover, in their book chapter [37], discussed and illustrated many blockchain techniques for electronic health records to ensure confidentiality, patient data protection, storage reduction space, and resistance to attacks. They also investigated the integration of blockchain in pharmaceutical and medical manufacturers. In addition, they presented the contribution of blockchain technology in biomedical and healthcare applications. Moreover, Kaddoura et al. [38] proposed a novel approach to detect malicious transactions in healthcare databases with minimal time. Based on database damage assessment, their proposed algorithm provided a low execution time; in addition, it enhanced memory allocation because infected transactions are quarantine.

2.4.1 IoT security in health informatics using machine learning

Gopalan et al. [39] provided the effectiveness of integrating artificial intelligence tools in protecting IoT healthcare networks against cyberattacks through research works during the past few years. Because attacks can take different forms, security also takes various forms: physical and information security. Physical security includes hardware, software, data, and network [40]. However, information security consists of various tools and techniques to detect and prevent threats [41]. Network, cloud, endpoint, and mobile securities are examples of Information security.

Throughout the past years, researchers have provided frameworks and artificial intelligence techniques to explore IoT security problems for different healthcare scenarios. For instance, many research works applied machine learning and deep learning methods to protect IoT networks to the different IoT layers, showing that existing security approaches need improvement for adequate security measures such as encryption, authentication, and application security [42]. A standard approach was proposed to analyze different cloud-based IoT healthcare systems using many machine learning approaches [43]. Their approach investigated information security issues through gathering and processing data from wearable sensors in healthcare systems. As for network security, machine learning and deep learning techniques were called to improve existing network security solutions [44]. The limitations of existing solutions include resource constraints, massive real-time data, dynamic behavior of the network, and heterogeneity.

Using different AI methods and algorithms, many works investigated IoT security issues to:

1. Increase the privacy of IoT in healthcare systems by using multischeme crypto deep neural networks [45]
2. Detect insider attacks by using Bayesian algorithm to evaluate the performance of healthcare environment [46]
3. Protect personal information privacy by using the *k*-means algorithm to enable clustering analysis for privacy preservation [47].

2.4.2 IoT security and COVID-19 pandemic

COVID-19 is a severe and fatal virus. The world health organization has declared it a global pandemic. Accordingly, the lifestyle of most organizations and individuals has changed. Most industrial and economic sectors have moved to work and cooperate through the Internet to keep their businesses running [48]. Thus, the IoT framework influenced all those sectors. In particular, the healthcare sector took advantage of integrating the IoT technology into its existing systems to fight the spread of this severe coronavirus. During the COVID-19 pandemic, workers and authorities in the health sector need to collect data to have the capability to manage and control the rapidly spreading of the pandemic virus. Consequently, the data serves to monitor the diagnosis of infection and track the spread of the virus within communities. Marais et al. [49] and Kobo et al. [50] adopted two technologies – LoRaWAN and SDWSN – to provide health workers tools to monitor the spread of the coronavirus in different locations and collect data for further analysis in the process of fighting the pandemic virus.

Sabukunze et al. [51] developed an IoT system design that monitors COVID-19 patients using sensors to measure the temperature, blood oxygen, and heart rate. The system will then alert the patient and the healthcare provider for any abnormal behavior in the patient's body. During monitoring and alerting stages, data are collected and sent to the healthcare IoT server for further analysis to reduce mortality, severe infection side effects, and hospital admission. Similarly, Paganelli et al. [52] proposed a complete IoT-based conceptual architecture that remotely monitors patients at home and in the hospital for better treatment. The requirements of their proposed architecture encounter scalability, reliability, privacy, security, interoperability, and network dynamics. Their study explained the mechanisms used for monitoring patients through wearable devices to ensure usefulness and flexibility for physicians and healthcare workers. Moreover, Dagliati et al. [53] proposed collaborative infrastructures and database sharing using health informatics. The IoT technology helps to share critical data (especially during the COVID-19 pandemic) to support day-to-day activities and provide the research professionals with the essential information to concentrate on studying drug manufacturing and treatments given for patients.

In-person medical services were not applicable during the COVID-19 pandemic because of lockdowns and hospital booming in patients. Consequently, telehealth intensively took place for nonclinical services. Due to better telecommunication technologies (video conferencing and 5G) development in recent years, doctors and healthcare professionals have used telehealth services to fight the current pandemic [54]. Thus, a need for any outside consultancy for a severe case will have treatment through telemedicine.

In-person medical services were not applicable during the COVID-19 pandemic because of lockdowns and hospital booming in patients. Consequently, telehealth intensively took place for nonclinical services [54]. This pandemic significantly impacted healthcare systems, including confidentiality, integrity, and availability [55]. Many breaches affected the healthcare organization workflow, such as unauthorized access to personal records, illegal view of patients' medical conditions, malicious deletion of data, loss of patients' records, etc. Accordingly, breaches had impacts on healthcare data, including misdiagnosis, stigmatization, denial of healthcare, and operational impacts.

Because cybersecurity is a common interest for all parties, protecting data from cyberattacks is the primary key to the effectiveness of this industry. Protecting healthcare data starts from detecting threats, recognizing their types, preventing them from reaching confidential data, and altering it. In the following section, supervised machine learning techniques that were applied in the literature were briefly shown to ensure the hard work done in this field.

2.5 Supervised machine learning techniques in detecting malicious behavior in health informatics

Because machine-learning techniques have proved advancements in multiple fields such as computer vision, speech recognition, text analysis, etc. The healthcare industry has started to integrate artificial intelligence techniques to assist physicians in examining patients, particularly after the cumulative stored data for patients and their cases [56]. In addition, the advancement in cloud computing, edge computing, big data, and mobile communication has increased the hunger in the healthcare systems to adopt technological and smart techniques. However, any technological advancement has its pros and cons; one of the critical disadvantages of involving technology in one of the most vital industries consists of the high risk of cyberattacks, including accessing, stealing, or altering patients' personal information [56]. Therefore, the integration of machine learning approaches helps detect and prevent attacks based on accurate prediction outcomes.

This section will introduce different literary works that applied machine learning techniques in the healthcare field to detect malicious behavior in the network. Most applied machine learning techniques in the healthcare field were related to some supervised machine learning models based on the literature. Very few studies were about unsupervised or combinations of unsupervised and supervised machine learning approaches. Deep learning techniques were also rarely used in previous literature. For this reason, the focus in this section is related to studying cybersecurity issues using supervised machine learning.

Researchers tackled health security using supervised machine learning from different perspectives. Attacks are detected from sensors and medical devices, spotted during the authentication and access control stage, and discovered as intrusion and malware. Table 2.1 shows references related to these three topics, showing the applied methods, and dataset used in each reference study.

2.5.1 Datasets experimental results

Using different types of datasets with different sizes (see Fig. 2.2), the authors in the papers listed in Table 2.1 have implemented their proposed models. They generated training and testing dataset samples extracted from the original datasets. It is essential to mention that some studies were done on actual collected data; however, other studies were based on running simulations under normal and suspicious network behavior. Note that some of the studies did not mention the exact size of their dataset used; thus, they were not included in the figure.

The datasets sizes vary between small (1000) dataset and large (150,000) dataset.

Table 2.1 Some reference papers for detecting malicious behavior in healthcare sensors, medical devices, authentication, and intrusion.

Paper no.	Paper	Detection type	Method(s)	Dataset
1	Gao and Thamilarasu [57]	Anomaly based attack detection	Support vector machine and *k*-means	Simulation data
2	Rathore et al. [58]	Anomaly false alarm	Multilayer perceptron and linear support vector machine	Diabetic dataset
3	Kintzlinger et al. [59]	Anomaly attack	Support vector machine	Own clinical data
4	Barros et al. [60]	Access control intrusion	Naïve Bayes, support vector machine, multilayer perceptron, random forest	Stress recognition database
5	Musale et al. [61]	Access control abnormal behavior	Random forest	Own dataset
6	Begli et al. [62]	Anomaly and signature-based intrusion detection	Support vector machine	NSL-KDD
7	Odesile and Thamilarasu [63]	Anomaly intrusion detection	K-nearest neighbor, support vector machine, decision tree, random forest	Own dataset
8	Fernandez Maimo et al. [64]	Signature-based and anomaly-based intrusion detection	Support vector machine	Own real-test dataset

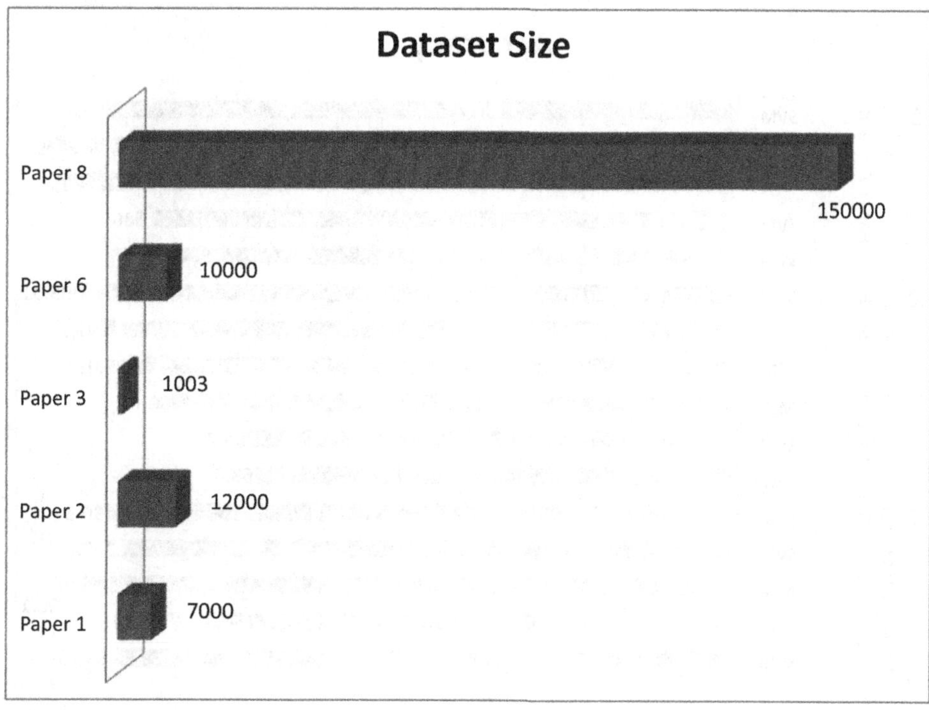

FIGURE 2.2

Dataset sizes (number of records/data collected) used in the papers studies represented in this chapter.

2.5.1.1 Accuracy measurement

The above studies measured the accuracy percentage for the dataset used using different supervised machine learning techniques, as shown in Fig. 2.3. Interestingly, the support vector machine algorithm is mainly used in these studies, showing variance results. For instance, in the stress recognition dataset (Paper 4), the accuracy is considerably low (68.6%) to that in NSL-KDD (Paper 6) (95.1%) or that of a real-time dataset (Paper 8) with a value of 90.4%.

The accuracy values vary between 68.6% and 99%.

2.5.1.2 False positive rate measurement

In Fig. 2.4, the false positive rate percentage is measured in some studies for different machine learning algorithms. It is shown that both RF and *k*-means have the highest false positive rate among the others with 9.97% and 9.48%, respectively.

The false positive rate values vary between 4.6% and 9.97%.

2.5.1.3 F1-score and precision measurements

The F1-score and Precision percentages are measured in two paper studies (Papers 1 and 8). In paper 1, it is clearly remarked that the multilayer perceptron algorithm can achieve better

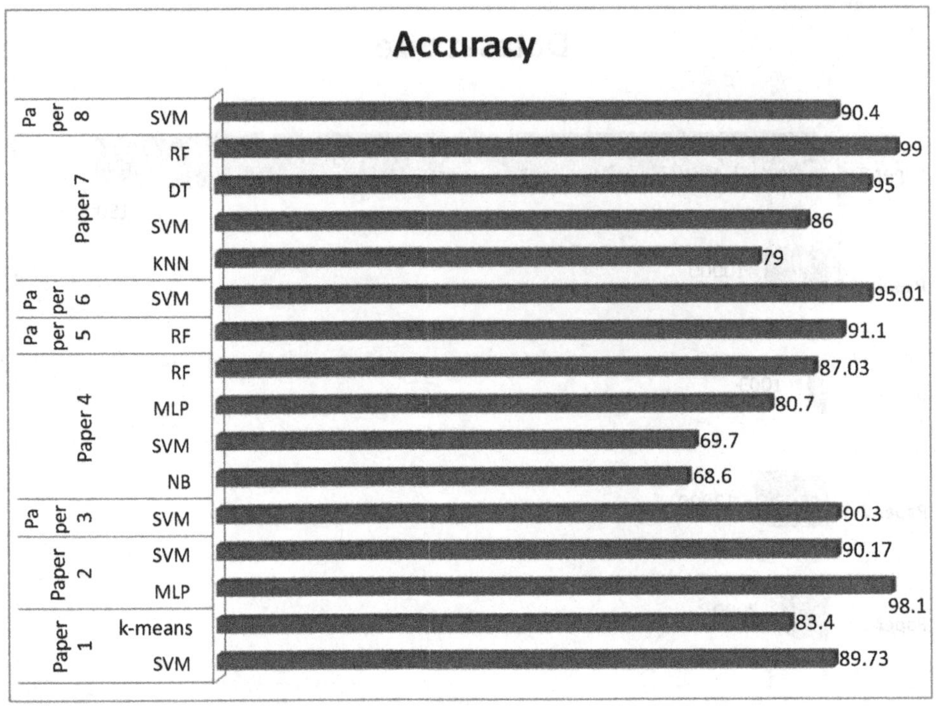

FIGURE 2.3

The accuracy percentage measurements achieved in the different studies presented in this chapter.

performance than the support vector machine algorithm for simulated data with a difference of around 11%. However, in a real-time dataset (Paper 8), the support vector machine algorithm performs well - F1-score: 95.96% and Precision: 92.32% (see Fig. 2.5).

2.5.2 Discussion

Some supervised machine learning techniques performed well in detecting different types of malicious behavior such as anomaly-based, false anomaly alarm, sensor, authentication/access control, and privacy attacks. However, these studies lack some essential parameters that should be estimated. Table 2.2 shows some of the parameters that, once considered, attacks detection may gradually increase.

From a scientific perspective, due to various limitations explored in previous works as shown in Table 2.2, researchers should be more involved in elaborating machine-learning techniques such as deep learning in the healthcare systems field because deep learning applications execute features by themselves [65]. A deep learning algorithm can scan data to determine specific features that are correlated and combined to provide a faster learning technique automatically.

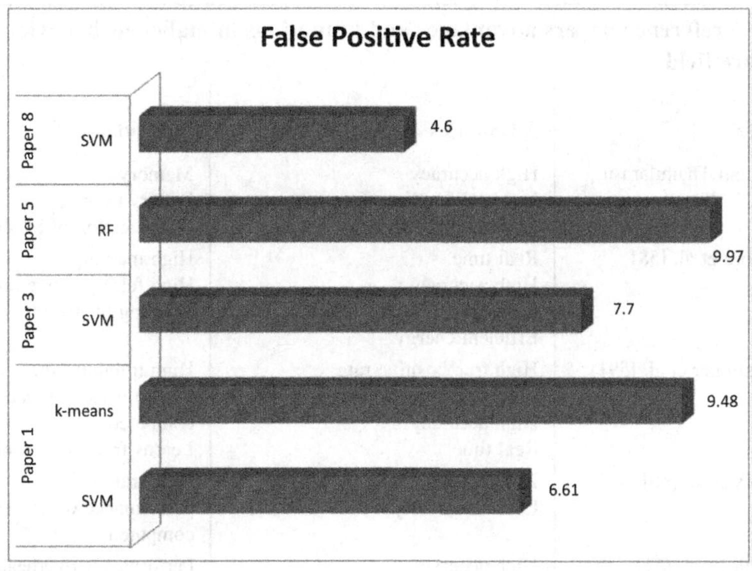

FIGURE 2.4

The false positive rate percentage measurements achieved in some studies presented in this chapter.

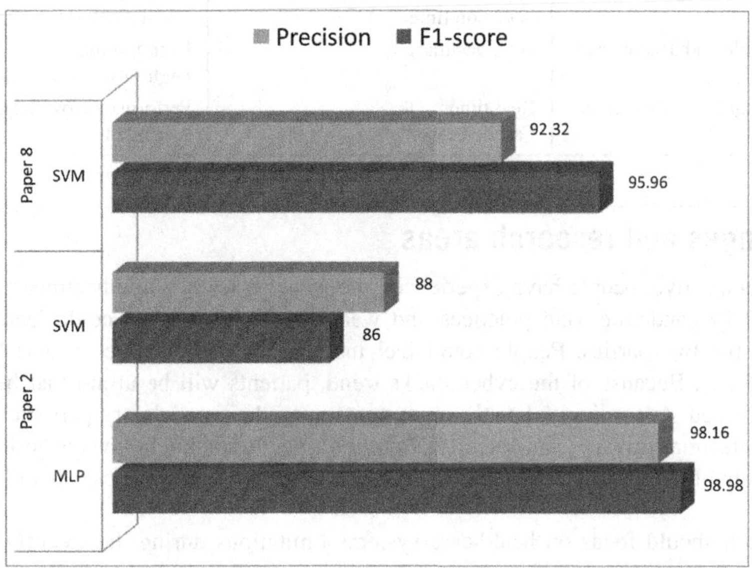

FIGURE 2.5

The f1-score and precision percentage measurements achieved in some studies presented in this chapter.

Table 2.2 Some reference papers advantages and limitations in malicious behavior detection in the healthcare field.

Paper no.	Paper	Advantages	Limitations
1	Gao and Thamilarasu [57]	High accuracy Low training time Low false positive rate	Memory Battery usage NO detection of insider attacks
2	Rathore et al. [58]	Real time High accuracy Reliable Efficient energy	High memory High false positive rate Training overhead
3	Kintzlinger et al. [59]	High true positive rate Low false positive rate High accuracy Real time	High training time Performance overhead not computed Learns from only benign data
4	Barros et al. [60]	High accuracy Low complexity	High training performance overhead not computed
5	Musale et al. [61]	High accuracy Lightweight	Performance overhead not computed Environment affects accuracy Authentication fails for smart home Accuracy decreases when increasing users
6	Begli et al. [62]	High detection convenient detection time	Memory overhead Old dataset
7	Odesile and Thamilarasu [63]	High accuracy	High training time High false positive rate
8	Fernandez Maimo et al. [64]	Real time Fast detection	Performance overhead not computed

2.6 Challenges and research areas

From an IoT perspective, people have experienced the weaknesses of some healthcare systems during the COVID-19 pandemic. Bad practices and weak health informatics could lead to trust problems between the two parties. People could feel insecure about their data misuse by healthcare organizations if any. Because of the cyberattacks trend, patients will be afraid that healthcare systems will be attacked. Accordingly, healthcare systems upgrade is mandatory periodically to collaborate in data collection, privacy, and security. Another aspect that should always be in mind is that information technologies are rapidly growing. Thus, health informatics should be updated periodically.

More research should focus on healthcare systems limitations during the COVID-19 pandemic to help increase awareness and performance of healthcare systems during such disastrous events in the future. Throughout the coming years, collecting data related to this pandemic after millions of people have been affected will be available. Thus, scientists will have the opportunity to elaborate more about this pandemic, better analyze its root causes and triggers. Accordingly, based on the advantages of deep over supervised/non supervised machine learning, new studies should show better performance in terms of accuracy and lower loss rate [66].

2.7 Conclusion

IoT is an advanced and endorsed technology that consists of complementary systems to healthcare systems. It facilitates patients' and physicians' communication, improves healthcare management, and supports patients monitoring remotely. Therefore, this chapter addressed the importance of IoT in organizations' workflow and individual contact. During the COVID-19 pandemic, the IoT technology helped to limit the spread of coronavirus. Thus, telemedicine service is one of the most helpful services provided by the IoT platform.

This chapter presented the current problems of healthcare systems with their corresponding suggested solutions in recent research, particularly in healthcare systems security. Many studies evaluated the involvement of supervised machine learning techniques in improving intrusion detection through the healthcare network, sensors, and medical devices. Different machine learning techniques show their advantages and limitations after being applied on various public, private, simulated datasets.

References

[1] S.K. Goudos, P.I. Dallas, S. Chatziefthymiou, S. Kyriazakos, A survey of IoT key enabling and future technologies: 5G, mobile IoT, sematic web and applications, Wirel. Personal. Commun. 97 (2) (2017) 1645–1675.

[2] S. Amendola, R. Lodato, S. Manzari, C. Occhiuzzi, G. Marrocco, RFID technology for IoT-based personal healthcare in smart spaces, IEEE Internet Things J. 1 (2) (2014) 144–152.

[3] L. Catarinucci, D. Donno de, L. Mainetti, L. Palano, L. Patrono, M.L. Stefanizzi, et al., An iot-aware architecture for smart healthcare systems, IEEE Internet Things J. 2 (6) (2015) 515–526.

[4] A. Edmondson, What hospitals overwhelmed by Covid-19 can learn from startups, 2020, May 22. Retrieved June 20, 2021, from Harvard Business Review. https://hbr.org/2020/05/what-hospitals-overwhelmed-by-covid-19-can-learn-from-startups.

[5] M. Jofre, D. Navarro-Llobet, R. Agulló, J. Puig, G. Gonzalez-Granadillo, J. Mora Zamorano, et al., Cybersecurity and privacy risk assessment of point-of-care systems in healthcare—a use case approach, Appl. Sci. 11 (15) (2021) 669.

[6] P. Gope, Y. Gheraibia, S. Kabir, B. Sikdar, A secure IoT-based modern healthcare system with fault-tolerant decision making process, IEEE J. Biomed. Health Inform. 25 (3) (2020) 862–873.

[7] R. Haraty, S. Kaddoura, A. Zekri, Transaction dependency based approach for database damage assessment using a matrix, Int. J. Semantic Web Inf. Syst. 13 (2) (2017) 74–86.

[8] A.J. Coronado, T.L. Wong, Healthcare cybersecurity risk management: keys to an effective plan, Biomed. Instrum. Technol. 48 (s1) (2014) 26–30.

[9] H. Tao, M.Z. Bhuiyan, A.N. Abdalla, M.M. Hassan, J.M. Zain, T. Hayajneh, Secured data collection with hardware-based ciphers for IoT-based healthcare, IEEE Internet Things J. 6 (1) (2019) 410–420.

[10] J. Li, J. Cai, F. Khan, A.U. Rehman, V. Balasubramaniam, J. Sun, et al., A secured framework for SDN-based edge computing in IoT-enabled healthcare system, IEEE Access. 8 (2020) 135479–135490.

[11] D. Serpanos, M. Wolf, Internet-of-Things (IoT) Systems: Architectures, Algorithms, Methodologies, Springer, 2017.

[12] G.J. Joyia, R.M. Liaqat, A. Farooq, S. Rehman, Internet of medical things (IOMT): applications, benefits and future challenges in healthcare domain, J. Commun. 12 (4) (2017) 240–247.

[13] S. Vishnu, S.J. Ramson, R. Jegan, Internet of medical things (IoMT)-an overview, in: 2020 5th International Conference on Devices, Circuits and Systems (ICDCS), 2020, pp. 101–104. IEEE.

[14] B. Abramiuc, S. Zinger, P.H. de With, N. de Vries-Farrouh, M.M. van Gilst, B. Bloem, et al., Home video monitoring system for neurodegenerative diseases based on commercial HD cameras, in: 2015 IEEE 5th International Conference on Consumer Electronics-Berlin (ICCE-Berlin), 2015, pp. 489–492. IEEE.

[15] A. Karakostas, G. Meditskos, T.G. Stavropoulos, I. Kompatsiaris, M. Tsolaki, A sensor-based framework to support clinicians in dementia assessment: the results of a pilot study, Ambient. Intelligence-Software Appl. (2015) 213–221.

[16] M. Ali, L. Albasha, H. Al-Nashash, A bluetooth low energy implantable glucose monitoring system, in: 2011 8th European Radar Conference, 2011, pp. 377–380. IEEE.

[17] R. Paoli, F.J. Fernández-Luque, G. Doménech, F. Martínez, J. Zapata, R. Ruiz, A system for ubiquitous fall monitoring at home via a wireless sensor network and a wearable mote, Expert. Syst. Appl. 39 (5) (2012) 5566–5575.

[18] S.H. Almotiri, M.A. Khan, M.A. Alghamdi, Mobile health (m-health) system in the context of IoT, in: 2016 IEEE 4th International Conference on Future Internet of Things and Cloud Workshops (FiCloudW), 2016, pp. 39–42. IEEE.

[19] S. Kaddoura, R.A. Haraty, A. Zekri, M. Masud, Tracking and repairing damaged healthcare databases using the matrix, Int. J. Distrib. Sens. Netw. 11 (11) (2015) 914305.

[20] L. Hadlington, Human factors in cybersecurity; examining the link between Internet addiction, impulsivity, attitudes towards cybersecurity, and risky cybersecurity behaviours, Heliyon 3 (7) (2017) e00346.

[21] S. Murphy, Is cybersecurity possible in healthcare, Natl Cybersecur. Inst. J. 1 (3) (2015) 49–63.

[22] Ö. Kafali, J. Jones, M. Petruso, L. Williams, M.P. Singh, How good is a security policy against real breaches? A HIPAA case study, in: 39th International Conference on Software Engineering (ICSE), 2017, pp. 530–540. IEEE/ACM.

[23] N.S. Abouzakhar, A. Jones, O. Angelopoulou, Internet of things security: a review of risks and threats to healthcare sector, in: IEEE International Conference on Internet of Things (iThings) and IEEE Green Computing and Communications (GreenCom) and IEEE Cyber, Phys. Soc. Comput. (CPSCom) IEEE Smart Data (SmartData), 2017, pp. 373–378. IEEE.

[24] N. Abouzakhar, Critical infrastructure cybersecurity: a review of recent threats and violations, A Treatise Electricity Magnetism 2 (2013) 68–73.

[25] R.L. Krutz, R.D. Vines, Cloud Security: A Comprehensive Guide to Secure Cloud Computing, Wiley Publishing, 2010.

[26] V. Winkler, Cloud computing: virtual cloud security, Technet Magazine, Microsoft (2012).

[27] V.J. Winkler, Securing the Cloud: Cloud Computer Security Techniques and Tactics, Elsevier, 2011.

[28] T.W. Reader, A. Gillespie, J. Roberts, Patient complaints in healthcare systems: a systematic review and coding taxonomy, BMJ Qual. Saf. 23 (8) (2014) 678–689.

[29] H. Sullivan, 7 Major challenges facing the healthcare industry in 2021, 2020. Retrieved June 20, 2021, from MailMyStatements. https://mailmystatements.com/2020/10/27/2019challenges/.

[30] A. Alamri, Ontology middleware for integration of IoT healthcare information systems in EHR systems, Computers 7 (4) (2018) 51.

[31] M.H. Alanazi, B. Soh, Behavioral intention to use IoT technology in healthcare settings, Eng. Technol. Appl. Sci. Res. 9 (5) (2019) 4769–4774.

[32] I.L. Pribadi, M. Suryanegara, Regulatory recommendations for IoT smart-health care services by using privacy impact assessment (PIA), in: 2017 15th International Conference on Quality in Research (QiR): International Symposium on Electrical and Computer Engineering, 2017, pp. 491–496. IEEE.

[33] T. Saheb, L. Izadi, Paradigm of IoT big data analytics in the healthcare industry: a review of scientific literature and mapping of research trends, Telemat. Inform. 41 (2019) 70–85.

[34] M. Quwaider, Y. Jararweh, A cloud supported model for efficient community health awareness, Pervasive Mob. Comput. 28 (2016) 35–50.

[35] A. Paul, H. Pinjari, W.H. Hong, H.C. Seo, S. Rho, Fog computing-based IoT for health monitoring system, J. Sens. (2018).

[36] M. Tahir, M. Sardaraz, S. Muhammad, M. Saud Khan, A lightweight authentication and authorization framework for blockchain-enabled IoT network in health-informatics, Sustainability 12 (17) (2020) 6960.

[37] S. Kaddoura, R. Grati, Blockchain for healthcare and medical systems, in: A.B. Mnaouer, L.C. Fourati (Eds.), Enabling Blockchain Technology for Secure Networking and Communications, IGI, 2021, pp. 249–270.

[38] S. Kaddoura, R.A. Haraty, K. Al Kontar, O. Alfandi, A parallelized database damage assessment approach after cyberattack for healthcare systems, Future Internet 13 (4) (2021) 90.

[39] S.S. Gopalan, A. Raza, W. Almobaideen, IoT security in healthcare using AI: a survey, in: International Conference on Communications, Signal Processing, and Their Applications, 2021, pp. 1–6. IEEE.

[40] M. Saadeh, A. Sleit, M. Qatawneh, W. Almobaideen, Authentication techniques for the internet of things: a survey, in: Cybersecurity and cyberforensics conference, 2016, pp. 28–34. IEEE.

[41] P.A. Williams, V. McCauley, Always connected: the security challenges of the healthcare Internet of Things, in: IEEE 3rd World Forum on Internet of Things, 2016, pp. 30–35. IEEE.

[42] M.A. Al-Garadi, A. Mohamed, A.K. Al-Ali, X. Du, I. Ali, M. Guizani, A survey of machine and deep learning methods for internet of things (IoT) security, IEEE Commun. Surv. & Tutor. 22 (3) (2020) 1646–1685.

[43] P. Ghosal, D. Das, I. Das, Extensive survey on cloud-based IoT-healthcare and security using machine learning, in: Fourth International Conference on Research in Computational Intelligence and Communication Networks (ICRCICN), 2018, pp. 1–5. IEEE.

[44] F. Hussain, R. Hussain, S.A. Hassan, E. Hossain, Machine learning in IoT security: current solutions and future challenges, IEEE Commun. Surv. Tutor. (2020) 1686–1721.

[45] O.A. Kwabena, Z. Qin, T. Zhuang, Z. Qin, Mscryptonet: multi-scheme privacy-preserving deep learning in cloud computing, IEEE Access. (2019) 29344–29354.

[46] W. Meng, K.K. Choo, S. Furnell, A.V. Vasilakos, C.W. Probst, Owards Bayesian-based trust management for insider attacks in healthcare software-defined networks, IEEE Trans. Netw. Serv. Manag. 15 (2) (2018) 761–773.

[47] Z. Guan, Z. Lv, X. Du, L. Wu, M. Guizani, Achieving data utility-privacy tradeoff in Internet of medical things: a machine learning approach, Future Gener. Computer Syst. 98 (1) (2019) 60–68.

[48] M. Yousif, C. Hewage, L. Nawaf, IoT technologies during and beyond COVID-19: a comprehensive review, Future Internet 13 (5) (2021) 105.

[49] J.M. Marais, A.M. Abu-Mahfouz, G.P. Hancke, A survey on the viability of confirmed traffic in a LoRaWAN, IEEE Access. 8 (2020) 9296–9311.

[50] H.I. Kobo, A.M. Abu-Mahfouz, G.P. Hancke, A survey on software-defined wireless sensor networks: challenges and design requirements, IEEE Access. 5 (2017) 1872–1899.

[51] I.D. Sabukunze, D.B. Setyohadi, M. Sulistyoningsih, Designing an IoT based smart monitoring and emergency alert system for Covid19 patients, in: 2021 6th International Conference for Convergence in Technology (I2CT), 2021, pp. 1–5. IEEE.

[52] A. Paganelli, P. Velmovitsky, P. Miranda, A. Branco, P. Alencar, D. Cowan, et al., A conceptual IoT-based early-warning architecture for remote monitoring of COVID-19 patients in wards and at home, Internet Things (2021). 100399.

[53] A. Dagliati, A. Malovini, V. Tibollo, R. Bellazzi, Health informatics and EHR to support clinical research in the COVID-19 pandemic: an overview, Brief. Bioinforma. 22 (2) (2021) 812–822.

[54] J. Ye, The role of health technology and informatics in a global public health emergency: practices and implications from the COVID-19 pandemic, JMIR Med. Inform. 8 (7) (2020) e19866.

[55] K. Okereafor, O. Adebola, Healthcare Cybersecurity Lessons from Covid, J. Homepage 9 (4) (2021). Available from: http://ijmr.net.

[56] A. Qayyum, J. Qadir, M. Bilal, A. Al-Fuqaha, Secure and robust machine learning for healthcare: a survey, IEEE Rev. Biomed. Eng. 14 (2020) 156–180.

[57] S. Gao, G. Thamilarasu, Machine-learning classifiers for security in connected medical devices, in: 26th International Conference on Computer Communication and Networks, 2017, pp. 1–5. IEEE.

[58] H. Rathore, L. Wenzel, A.K. Al-Ali, A. Mohamed, X. Du, M. Guizani, Multi-layer perceptron model on chip for secure diabetic treatment, IEEE Access. 6 (2018) 44718–44730.

[59] M. Kintzlinger, A. Cohen, N. Nissim, M. Rav-Acha, V. Khalameizer, Y. Elovici, et al., CardiWall: a trusted firewall for the detection of malicious clinical programming of cardiac implantable electronic devices, IEEE Access. 8 (2020) 48123–48140.

[60] A. Barros, D. Rosário, P. Resque, E. Cerqueira, Heart of IoT: ECG as biometric sign for authentication and identification. in: 15th International Wireless Communications & Mobile Computing Conference (IWCMC), 2019, pp. 307–312. IEEE.

[61] P. Musale, D. Baek, N. Werellagama, S.S. Woo, B.J. Choi, You walk, we authenticate: lightweight seamless authentication based on gait in wearable IoT systems, IEEE Access. 7 (2019) 37883–37895.

[62] M. Begli, F. Derakhshan, H. Karimipour, A layered intrusion detection system for critical infrastructure using machine learning, in: 7th International Conference on Smart Energy Grid Engineering (SEGE), 2019, pp. 120–124. IEEE.

[63] A. Odesile, G. Thamilarasu, Distributed intrusion detection using mobile agents in wireless body area networks, in: 7th International Conference on Emerging Security Technologies (EST), 2017, pp. 144–149. IEEE.

[64] L. Fernandez Maimo, A. Huertas Celdran, A.L. Perales Gomez, F.J. Garcia Clemente, J. Weimer, I. Lee, Intelligent and dynamic ransomware spread detection and mitigation in integrated clinical environments, Sensors 1114 (2019).

[65] C. Shen, D. Bguyen, Z. Zhou, S. Jiang, B. Dong, X. Jia, An introduction to deep learning in medical physics: advantages, potential, and challenges, Phys. Med. Biol. 65 (5) (2020). 05TR01.

[66] A. Agrawal, A. Choudhary, Deep materials informatics: applications of deep learning in materials science, MRS Commun. 9 (3) (2019) 779–792.

CHAPTER

3

Design of a novel privacy preservation based cyber security system framework for secure medical data transactions in cloud storage

S.K.B. Sangeetha[1], K. Veningston[2], Vanlin Sathya[3] and R. Kanthavel[4]

[1]Department of CSE, SRM Institute of Science and Technology, Vadapalani Campus, Chennai, Tamil Nadu, India [2]Department of CSE, National Institute of Technology, Sri Nagar, Jammu and Kashmir, India [3]System Engineer, Celona Inc., Cupertino, CA, United States [4]Department of CSE, King Khalid University, Kingdom of Saudi Arabia

3.1 Introduction

IoT has advanced greatly in the domain of processing large volumes of data since the introduction of cloud-based devices [1]. IoT provides a wide range of field applications for continuous monitoring in various industries, including the healthcare system, which is one of the IoT's new cloud users. The ability to effectively communicate, store, and handle huge volumes of data exchanged between devices, particularly healthcare data containing a large amount of patient data, is a challenge for IoT-based healthcare systems. Because of the ever-increasing population, the healthcare business must be able to efficiently manage people's health, which is made possible by implementing IoT models [2,3]. The purpose of the launch of a new cloud is to address data-related challenges in IoT healthcare systems [4].

Although the cloud provides a quick, flexible, adaptable, scalable, and cost-effective platform for the support and advancement of the IoT-based healthcare system, the privacy of extremely sensitive and vital patient data becomes a serious issue with the integration of the cloud with the healthcare system [5,6]. In the IoT-cloud-based healthcare system, data privacy and overhead processing are two important challenges. Thus diagnosing patients on the basis of data collected from IoT devices is a tricky aspect of healthcare systems. Several academics, however, have proposed alternative methods for evaluating the efficacy of cryptographic and noncryptographic data privacy measures [7,8].

There is a need to develop and improve the preservation of privacy methods since preserving sensitive data remains a difficult task for many organizations [9,10]. A large amount of unique or secret data have been obtained to examine the issue of privacy protection of data in healthcare organizations. A variety of solutions must be implemented to resolve privacy and accuracy issues while transmitting sensitive data stored in the cloud. As a means of tackling this difficult

Intelligent Edge Computing for Cyber Physical Applications. DOI: https://doi.org/10.1016/B978-0-323-99412-5.00006-X

predicament, intelligent solutions have promise. Medical data that must be processed and delivered in a cloud environment in a secure way is considered in this study.

Cyberattacks continue to evolve and grow more widely and regularly as is evident from backup and restore. Attackers create ideas and tactics faster than defense teams can keep up, and their tactics are becoming more sophisticated every year. Medical businesses are facing increasing strain from cyberattacks, which can take a variety of forms and affect a wide range of networks [1]. In the fight against cybercrime, medical organizations should make sure that their staff and customers are constantly aware of cyber threats, such as phishing attempts and social engineering. The fundamental issue confronting privacy and protection is the existence of multiple options for cloud providers to target personal data, resulting in a twofold increase in the user's anxiety about the data stored. While there are different data hiding tactics for various applications in the cloud area using big data, the vulnerabilities should be examined and fixed in future works [11–13].

This study proposes a novel cyber security technique called privacy preserved cyber security approach (PPCSA) that preserves privacy while storing and transporting medical data on the cloud. In the transformation of the solution, the key "*K*" is translated into another form using the Kronecker method in the first stage of the key generation process. In addition, row-wise duplication is used to generate the recovered key matrix. The proposed system model is described in Section 3.2, the experimental results are presented in Section 3.3, followed by conclusion in Section 3.4.

3.2 System model

Two important processes are included in the proposed privacy protection paradigm: (1) data sanitization and (2) data restoration. The sanitization procedure obscures confidential messages, preventing sensitive data from leaking to the cloud, which stores all natural and sensitive data [14]. A new privacy preserved cyber security approach is developed for the sanitization approach, which uses a key based on the Lion Algorithm (LA). The same key should be used during data restoration to effectively restore the original data. The process of sanitization refers to the pattern that covers confidential cloud data information. According to the present definition, a key is utilized to conduct the sanitization procedure, which can protect sensitive data. The proposed model's main objective is to create a key using an upgraded LA. If the approved individual uses the same key, the original data may be retrievable [15,16] (Fig. 3.1).

3.2.1 Objective and function of proposed model

The major goal of developing the best key for the sanitization process is to balance three parameters: Euclidean distance, hiding rate, and the rate at which information is preserved [17]. With the multiobjective function, the proposed privacy preservation method will efficiently execute the sanitization and restoration of cloud accounting transactions taken into account in this research effort.

The key extraction is illustrated with three parameters in Eq. (3.1).

$$O = w.(S + (1 - P_R)) \tag{3.1}$$

where w is the weight function, S is sanitized data adjustment, P_R is information preservation rate.

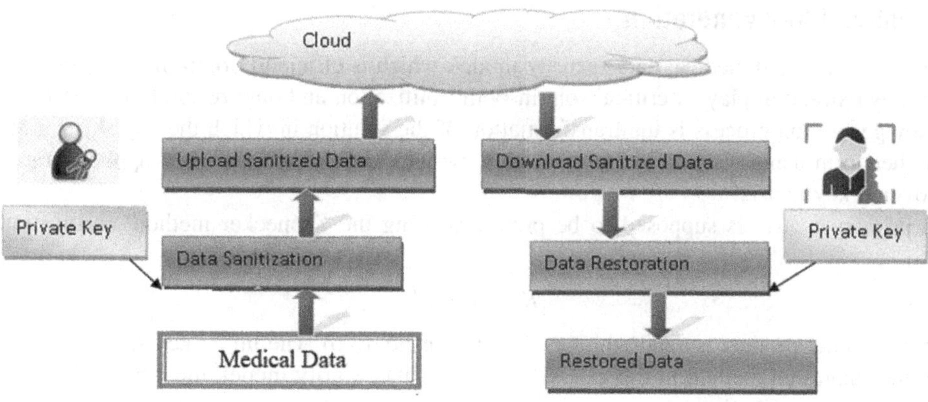

FIGURE 3.1

Proposed privacy preservation architecture.

Eq. (3.2) represents the formulation for the degree of modification by finding the Euclidean distance between the original dataset d and the sanitized dataset d'.

$$S = \text{EucDis}(d - d') \tag{3.2}$$

The percentage of sensitive objects in d' that are appropriately covered is known as the hiding rate h. Eq. (3.3) contains the mathematical method for computing the hiding rate. o' signifies sanitized objects or laws, while s denotes sensitive rules.

$$h = \frac{|o' \cap s|}{|s|} \tag{3.3}$$

As indicated in Eq. (3.4), the information preservation rate p is the inverse of the rate of information loss l.

$$p = 1 - \frac{|l - l'|}{|l|} \tag{3.4}$$

3.2.2 Sanitization process

Sanitization is a technique for masking sensitive medical information. The data is first converted to binary format and then the XOR procedure is applied to obtain the sanitized information. The sanitized data description from the original data and derived key is shown in Eq. (3.5).

$$d' = d \oplus K \tag{3.5}$$

where d' is sanitized data, d is original data, K is the optimal key.

The items are extracted after sanitization and are examined before sanitization is carried out. The sensitive laws, which are imported into the cloud, are then concealed from the sanitization process, boosting the security efficiency in the cloud domain. As a consequence, without any cyberattacks, the data would be secure for wider exposure.

3.2.3 Optimal key generation

Based on the suggested medical data security model, which is efficiently optimized by the proposed method, key extraction plays a critical role in both sanitization and data restoration. The first stage in the key generation process is the transformation of the solution in which the key, K, is translated into another form using the Kronecker method. Row-wise duplication has been applied to generate the recovered key matrix.

The primary matrix is supposed to be produced using the Kronecker method, as illustrated in Eq. (3.6).

$$K'' = K' \oplus K' \tag{3.6}$$

The Kronecker product is denoted by the symbol in Eq. (3.6). The initial key K, which is dependent on the suggested method in the existing medical data security model, must be calibrated properly as the main contribution.

3.2.4 Restoration process

The data is first converted to binary format. The proposed system conducts binary conversion to the determined optimal key. The same optimal key in the sanitization process is used for data restoration. Data restoration is performed based on Eq. (3.7), where the sanitized data is indicated as d' and the optimal key is K.

$$\alpha = d' \oplus K \tag{3.7}$$

3.2.5 Proposed privacy preserved cyber security approach

The proposed PPCSA, which is the modified form of standard LA, is the primary optimization for medical data protection.

The basic steps of the process are briefly outlined:

Step 1: Initialize the vector elements of V^{male}, V^{fem}, and V^{nor} ranging in between the limits when a > 1 in which $a = 1, 2, 3 \ldots i$.

Step 2: Calculate length of the lion $l = \{\text{if } m > 1 \text{ then m else } n\}$.

Step 3: If fitness(V^{male}) > fitness(r) then increase laggardness rate lr by one; if $lr > \max(lr)$ then territorial defense occurs.

Step 4: Update V^{fem+} as V^{fem} until the female generation count c reaches to maximum.

Step 5: Crossover and mutation, and the supplementary stage is gender clustering.

Step 6: The program uses territory defense to avoid the local optimal point and find other fitness comparable solutions.

When the criteria of Eq. (3.8) is met, the value of t' is chosen.

$$\text{fitness}(t') < \text{fitness}\left(V^{male}\right) \tag{3.8}$$

Step 7: Generation is denoted by g and threshold error is indicated by e.

Step 8: To repeat the process go to 4, otherwise end the process.

A few adjustments are included in the process crossover to improve the efficiency of classic LA. Instead of using the conventional crossover operation, the suggested technique uses a new crossover operation based on a random number r. This ideally constructed key is therefore successful in performing both sanitization and restoration in maintaining medical data and implementing cyber protection in the cloud environment.

3.3 Results and discussions

3.3.1 Experimental setup

The implementation of the proposed medical data privacy preservation model in the cloud sector for cyber security was done using Matlab, and its performance was evaluated. The rating is based on three datasets: WHO, CDC, and data.gov. Heart disease and diabetes risk can be predicted using the WHO repository dataset. The collection starts with 400 people, each of whom contains 67 factors that can be used to predict heart disease. The CDC dataset includes 1450 samples with 12 diabetes data characteristics. The number of iterations and the total number of solutions to validate the output of the proposed privacy preservation based key generation was compared with Lion Algorithm (LA), Whale Optimization Algorithm (WOA), and Gray Wolf Optimizer (GWO). Convergence study, statistical analysis, and the effect of chosen-plaintext attack (CPA) and known-plaintext attack (KPA) were performed during the performance validation process.

3.3.2 Performance metrics

The convergence analysis is shown in Fig. 3.2. For the remaining iterations in archiving the optimal convergence to reduce the objective function, the proposed PPCSA produced better results. The cost function of the implemented PPCSA is 25% better than WOA, 61% better than GWO, and 52% better than LA for the 100th iteration in Fig. 3.2. The suggested PPCSA displays its efficacy not just in these iterations, but also in subsequent iterations.

As demonstrated in Fig. 3.2, the given PPCSA is 20% greater than LA at the 100th iteration, 15% higher than GWO, and 18% higher than WOA. As a result, the anticipated sanitization of data and the restoration of medical data are deemed to be successful. The statistics of maximum execution in terms of best output, worst performance, mean, median, and standard deviations for three medical datasets are computed. For the three datasets, the proposed PPCSA outperforms the current WOA, GWO, and standard LA. The mean score is the total of the best and worst executions, the median is the point in the middle of the best and worst executions, and the standard deviation is the difference in the five executions.

For the proposed PPCSA, the worst instances are the minimum. In Fig. 3.3 for dataset 1. GWO, WOA, and LA are outperformed by 91%, 58%, and 51%, respectively. PPCSA sanitization has a lower average performance than GWO, WOA, and LA, with an average performance of 82% better than GWO, 93% better than WOA, and 8% better than LA. For the statistical analysis, the provided PPCSA produces a lower mean performance. The developed PPCSA has the lowest average compared with other algorithms, as shown in Fig. 3.3 for dataset 3. The used PPCSA's median number

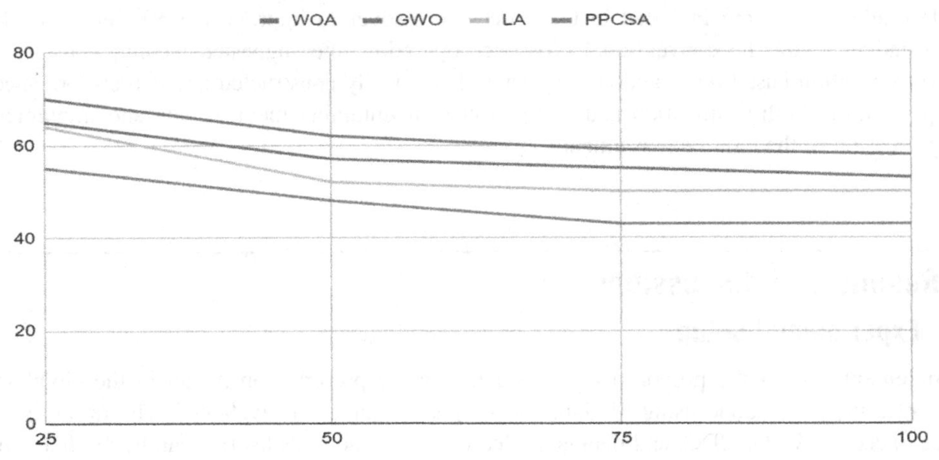

FIGURE 3.2

Convergence analysis.

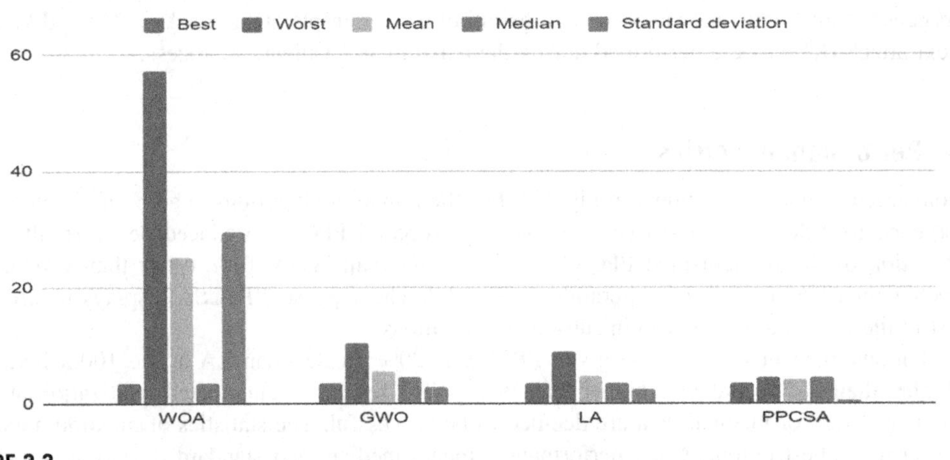

FIGURE 3.3

Statistical analysis.

is likewise lower. The three tables show that the proposed PPCSA is capable of both sanitizing and recovering cloud medical data.

CPA is a cryptanalysis attack model in which it is assumed that the attacker can obtain the ciphertexts for any plaintext. The link between the original and restored data once the CPA is completed. By experimenting with three test situations, the proposed PPCSA generates excellent efficiency in comparison to GWO, WOA, and standard LA. Fig. 3.4 shows that the dataset 1 PPCSA is 0.1% better than

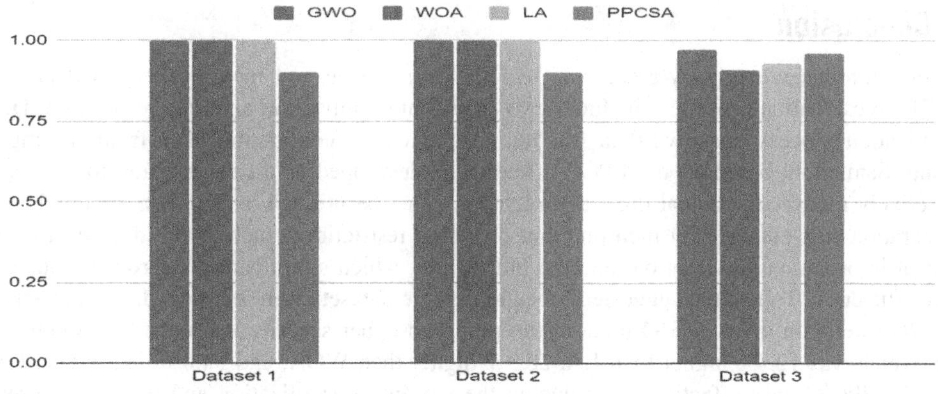

FIGURE 3.4

Chosen-plaintext attack analysis.

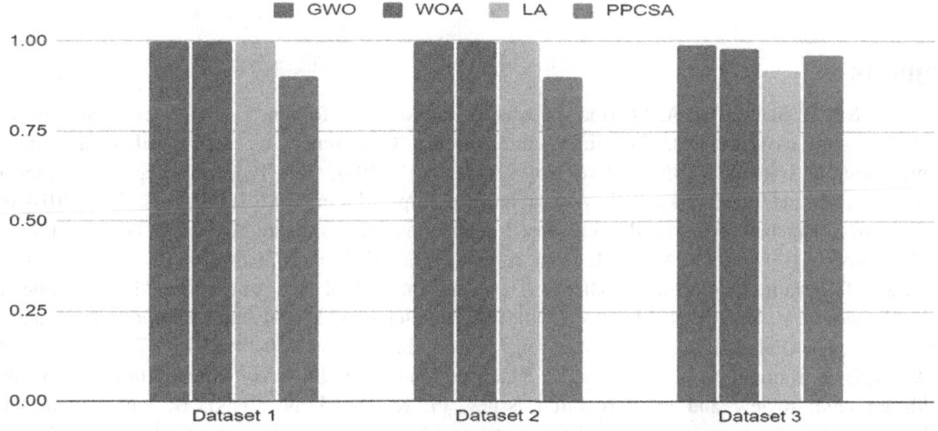

FIGURE 3.5

Known-plaintext attack analysis.

GWO, WOA, and LA. When the correlation is tested, the presented PPCSA of dataset 2 outperforms GWO, WOA, and LA by 0.2% (Fig. 3.5).

KPA is a type of cryptanalysis attack in which the attacker has access to both the plaintext and the encrypted version of the data. To validate the relationship between the original and restored data, the method is hacked with KPA connect. For dataset 1, the suggested PPCSA is 0.1% greater than GWO, WOA, and LA; however, for dataset 2, it is 0.01% higher than GWO, WOA, and LA. For dataset 3, the implemented PPCSA correlates 0.3% better than GWO, WOA, and LA. As a result of the foregoing findings, the suggested PPCSA solution can be determined to be suitable for cloud cyber security adoption.

3.4 Conclusion

This work introduces a privacy-preserving technique in storing and moving medical data on the cloud. The preservation procedure includes two steps: data sanitization and data restoration. During the sanitization process, sensitive data was buried, preventing confidential data from seeping into the cloud. Sanitation, based on the PPCSA, had to be developed as a new addition to the project. To effectively recover the actual data, a similar key must be utilized in data restoration. The best key generation was obtained by incorporating objective restrictions, such as the degree of change, the rate of information preservation, and the hiding rate, which significantly improved cloud cyber security efficiency. To test the suggested procedure, three datasets were employed. As a result, during the 76th iteration of the WHO dataset, the proposed cyber security approach had a better cost feature, which was 12.3% higher than LA, 28.9% higher than WOA, and 34.6% higher than GWO. Finally, the PPCSA was effective in producing the key in the sanitization and restoration process, which aided the protection of medical data without leaking any data to the cloud. The proposed method can also be used with any data that needs security.

References

[1] R. Dhaya, S.K.B. Sangeetha, A. Sharma, Jagadeesh, Improved performance of two server architecture in multiple client environments, in: IEEE International Conference on Advanced Computing and Communication Systems, ISBN XPlore No: 978-1-5090-4559-4, Shri Eswar College of Engineering, Coimbatore, 6th and 7th January 2017. Available from: https://doi.org/10.1109/ICACCS.2017.8014560.

[2] B. Pradhan, S. Bhattacharyya, K. Pal, IoT-based applications in healthcare devices, J. Healthc. Eng. 2021 (2021) 18Article ID 6632599. Available from: https://doi.org/10.1155/2021/6632599.

[3] S. Ansari, T. Aslam, J. Poncela, P. Otero, A. Ansari, Internet of things-based healthcare applications, in: B. Chowdhry, F. Shaikh, N. Mahoto (Eds.), IoT Architectures, Models, and Platforms for Smart City Applications, IGI Global, 2020, pp. 1–28. Available from: http://doi.org/10.4018/978-1-7998-1253-1.ch001.

[4] Z. Alansari, S. Soomro, M.R. Belgaum, S. Shamshirband, The rise of internet of things (IoT) in big healthcare data: review and open research issues, in: K. Saeed, N. Chaki, B. Pati, S. Bakshi, D. Mohapatra (Eds.), Progress in Advanced Computing and Intelligent Engineering. Advances in Intelligent Systems and Computing, 564, Springer, Singapore, 2018. Available from: https://doi.org/10.1007/978-981-10-6875-1_66.

[5] S. Deokar, M. Mangla, R. Akhare, A secure fog computing architecture for continuous health monitoring, in: S. Tanwar (Ed.), Fog Computing for Healthcare 4.0 Environments. Signals and Communication Technology, Springer, Cham, 2021. Available from: https://doi.org/10.1007/978-3-030-46197-3_11.

[6] W. Li, Y. Chai, F. Khan, et al., A comprehensive survey on machine learning-based big data analytics for IoT-enabled smart healthcare system, Mob. Netw. Appl. 26 (2021) 234–252. Available from: https://doi.org/10.1007/s11036-020-01700-6.

[7] S.K.B. Sangeetha, R. Dhaya, Deep learning era for future 6G wireless communications—theory, applications, and challenges, Artif. Intell. Tech. Wirel. Commun. Netw. (2022) 105–119. Available from: https://doi.org/10.1002/9781119821809.ch8.

[8] A. Ari, O. Ngangmo, C. Titouna, O. Thiare, Kolyang, A. Mohamadou, et al., Enabling privacy and security in cloud of things: architecture, applications, security & privacy challenges, Appl. Comput. Inform. (2019). Available from: https://doi.org/10.1016/j.aci.2019.11.005.

[9] A. Chaudhari, R. Bansode, Survey on securing IoT data using homomorphic encryption scheme, Int. J. Eng. Adv. Technol. 10 (2021) 76–81. Available from: https://doi.org/10.35940/ijeat.D2333.0410421.

[10] J.G. Pandey, C. Mitharwal, A. Karmakar, An RNS implementation of the elliptic curve cryptography for IoT security, in: 2019 First IEEE International Conference on Trust, Privacy and Security in Intelligent Systems and Applications (TPS-ISA), 2019, pp. 66–72. https://doi.org/10.1109/TPS-ISA48467.2019.00017.

[11] F. Farhin, M.S. Kaiser, M. Mahmud, Secured smart healthcare system: blockchain and bayesian inference based approach, in: M.S. Kaiser, A. Bandyopadhyay, M. Mahmud, K. Ray (Eds.), Proceedings of International Conference on Trends in Computational and Cognitive Engineering. Advances in Intelligent Systems and Computing, 1309, Springer, Singapore, 2021. Available from: https://doi.org/10.1007/978-981-33-4673-4_36.

[12] S. Selvaraj, S. Sundaravaradhan, Challenges and opportunities in IoT healthcare systems: a systematic review, SN Appl. Sci. 2 (2020) 139. Available from: https://doi.org/10.1007/s42452-019-1925-y.

[13] A.K. Singh, A. Anand, Z. Lv, H. Ko, A. Mohan, A survey on healthcare data: a security perspective, ACM Trans. Multimedia Comput. Commun. Appl. 17 (2021). Available from: https://doi.org/10.1145/3422816. 2s, Article 59 (June 2021), 26 p.

[14] O.I. Khalaf, K.A. Ogudo, S.K.B. Sangeetha, Design of graph-based layered learning-driven model for anomaly detection in distributed cloud IoT network, in: Mobile Information Systems, 2022, Article ID 6750757, 9 p. https://doi.org/10.1155/2022/6750757.

[15] S.K.B. Sangeetha, R. Dhaya, T. Dhruv, R. Shah, K. Dharanidharan, S.R. Praneeth, An empirical analysis of machine learning frameworks digital pathology in medical science, J. Phys. Conf. Ser. 1767 (2021) 012031. Available from: https://doi.org/10.1088/1742-6596/1767/1/012031.

[16] Y.I.N. Yuehong, The internet of things in healthcare: an overview, J. Ind. Inf. Integr. 1 (2016) 3–13. Available from: https://doi.org/10.1016/j.jii.2016.03.004. ISSN 2452-414X.

[17] W.T. Song, B. Hu, X.F. Zhao, Privacy protection of IoT based on fully homomorphic encryption. Wireless Commun. Mobile Comput. 2018 (2018). Article ID 5787930. https://doi.org/10.1155/2018/5787930.

CHAPTER

IoT-based BIM integrated model for energy and water management in smart homes

4

Mervin Ealiyas Mathews[1], Anandu E. Shaji[2], N. Anand[3], A. Diana Andrushia[4], Siew Choo Chin[5] and Eva Lubloy[6]

[1]*L&T Edutech, Larsen and Toubro Limited, Chennai, Tamil Nadu, India* [2]*Department of Civil Engineering, Mar Baselios Christian College of Engineering and Technology, Kuttikanam, Kerala, India* [3]*Department of Civil Engineering, Karunya Institute of Technology and Sciences, Coimbatore, Tamil Nadu, India* [4]*Department of Electronics & Communication Engineering, Karunya Institute of Technology and Sciences, Coimbatore, Tamil Nadu, India* [5]*Department of Civil Engineering, College of Engineering, Universiti Malaysia Pahang, Pekan, Pahang, Malaysia* [6]*Department of Construction Materials and Engineering Geology, Budapest University of Technology and Economics, Budapest, Hungary*

4.1 Introduction

Smart home is a concept that mainly focuses on enhancing the comfort of occupants and facilitating household activities [1]. Smart homes can be improved with Information and Communication Technologies (ICT) to provide the user with context-aware automated or assistive services in the form of ambient intelligence, remote home control, or home automation. The idea of "smart" homes is to integrate smartness into homes to guarantee residents' convenience, safety, and security while conserving energy. The smart home is commonly addressed as a smart house, home automation, intelligent home, adaptive home, or warehouse [2].

The first definition of a smart home was provided by Lutolf [3] in 1992. The idea of a smart home is the integration of different services in a home by using a common communication system. This system provides an economical, secure, comfortable, and energy-efficient operation of the home, which includes a high degree of intelligent functionality and flexibility. It also uses internet-connected devices to enable the remote monitoring and management of appliances and systems, such as lighting and heating.

Energy and water management are the key elements in a smart home, and a smart home functions to reduce energy consumption and manage water in and around the home effectively. According to Lashkari et al. [1], energy consumption has increased due to the rising population and expanding economy. With the improved quality of life, energy consumption will continue to increase, and the increment rates are expected to continue. High energy consumption will lead to high emissions of greenhouse gasses (GHG), which has a serious impact on the global environment. In 2004, the building energy usage in the European Union (EU) was 37% of the final energy, which is higher than other sectors such as industries (28%) and transport (32%). Unlike other sectors,

Intelligent Edge Computing for Cyber Physical Applications. DOI: https://doi.org/10.1016/B978-0-323-99412-5.00009-5

there are various attractive opportunities for buildings to reduce energy consumption to have lower costs and higher returns. Meanwhile, energy consumption in the residential sector in the year 2007 represented 21% of the total US demand. After the initial increase from 17% to 20% from the year 1950 to 1960, the amount has remained between 19% and 22% to date [4]. A huge amount of home energy is consumed ineffectively. It is reported that residential buildings are responsible for 40% of global power consumption [5]. The poor technology of energy management systems is the main cause of energy waste in homes.

Sustainable development goals (SDG) aim to ensure affordable, reliable, and sustainable modern energy for all occupants. Sustainable development requirements and the significant increase in energy costs necessitate the reduction in energy consumption without compromising on the comfort of the consumers. Integration of intelligent management systems in buildings will lead to less consumption of energy and save cost at the same time. The concept of smart homes has therefore received overwhelming attention in the last decade due to its potential in providing comfort to the occupants along with energy management.

On the other hand, the water industry is facing new challenges in managing sustainable urban water systems. External factors such as the impact of climate change, drought, and population growth in urban centers have increased the responsibility to adopt more sustainable management of the water sector [6]. Some of the main challenges in water management include cost coverage, monitoring of nonrevenue water (NRW), and knowledge of customers' demand for fairness in revenue [7]. It is important to have proper water management due to the growth in population and concentration of water needs. It is therefore necessary to use advanced technologies and the adoption of more robust management models to meet water demands [8].

Water stress has become a major issue due to the scarcity of freshwater reserves in different regions of the world. In 2025, it is estimated that almost half of the urban population will live in the water-stressed area as this precious source is becoming scarce rapidly [9–12]. According to studies, a possible solution to avoid a worldwide water crisis is by adopting water management and control systems based on automated solutions such as the IoT. The integration of the IoT methods is considered one of the best possible solutions that would enable the maintenance of a sustainable and cost-effective water supply [10,13–15]. Some of the advantages of water management in smart homes integrated with IoT technology include a better understanding of the water system, detection of leaks, conservation, and monitoring of water quality [16].

In the past three decades, a revolutionary approach like BIM has been developing in the field of construction and design. Xu et al. [17] defined BIM as "a model-based process of generating and managing coordinated and consistent building data that facilitates the accomplishment of established sustainable goals." This signifies that BIM has reached a level to facilitate high-level analysis as well as evaluations for buildings. These can be performed through techniques such as acoustic analysis, carbon emission, construction and demolition, waste management, operational energy, and water use. Additionally, BIM could be expressed in a 3D model of multidisciplinary data for various analyses [18]. The innovative development of BIM could provide opportunities to support green buildings through the application of high-tech programs or devices such as the IoT and smart devices.

In the context of smart homes, IoT refers to the nature of the interconnection of sensing and actuating devices and the ability to share information through a unified framework that could develop a common operating picture to enable innovative applications [19]. The IoT is an architecture that uses intelligent devices, smart mobile devices, single-board computers, different types of sensors, and actuators [20]. The critical importance of IoT lies in allowing sensitive technologies such as sensing, identification, and recognition run by advanced hardware, software, and cloud platforms to facilitate renovations and/or demolitions [21].

BIM-IoT integration is still in the early stage of development, which requires serious efforts to achieve a good understanding of the current usage [22]. The advantage of the real-time integration of the environmental and localization data will be help in operational construction and facility management by applying the cloud-based BIM platform [23].

4.1.1 Research objectives

The growing infrastructure necessitates the need for optimizing the energy demand in buildings. Many key influencing factors affect the energy performance of buildings, which include the shape and size of the building, orientation, window-to-wall ratio, and material used for wall and roof construction. Guidelines are suggested in NBC, ASHRAE, and Indian Green Building Council (IGBC) on how to model and design buildings that are more sustainable in saving energy [24–26]. A detailed analysis of how to optimize energy demand in buildings is discussed in this chapter.

The proposed study presents the energy analysis and modeling of a residential building based on Insight 360 modules. The analysis involves a sunlight exposure study, lighting efficiency, and window shade analysis to obtain the modified energy requirement. Many trials have been attempted to obtain the optimized energy by varying the design parameters that affect energy consumption. In addition, a plumbing scheme is modeled and proposed for optimum water usage. The results are presented with a detailed discussion.

The objective of this research is to disclose the effective usage of the IoT-based BIM Integrated model for energy and water management in smart homes. This includes energy analysis for energy management using the BIM process. In terms of water management systems in smart homes, a bi-directional plumbing system is proposed and discussed. The outcome of this study provides a clear understanding and knowledge on the integration of IoT and BIM in managing energy and water effectively in smart homes.

4.2 Background

The construction industry faces a lot of challenges in terms of material management, project coordination, cost control, and time management. Lack of planning, coordination, decision-making, monitoring, and integrated project delivery are some of the problems the industry undergoes during the construction phase. Among other industries, the building sector consumes 30% raw material and 42% of energy worldwide. Also, 37% of the material used for construction become waste and 38% of carbon emission are from buildings during its service period (CMAA report).

The proposed model of the building is considered in the location of Elappara, India, as shown in Fig. 4.1. Elappara is a village in the Idukki region of the state of Kerala. It covers an area of 91.85 km^2. The average temperature in Elappara varies between 18°C and 20°C. The precipitation rate is 2%, humidity rate is 85%, and wind velocity rate is 2 km/h. The region is located at

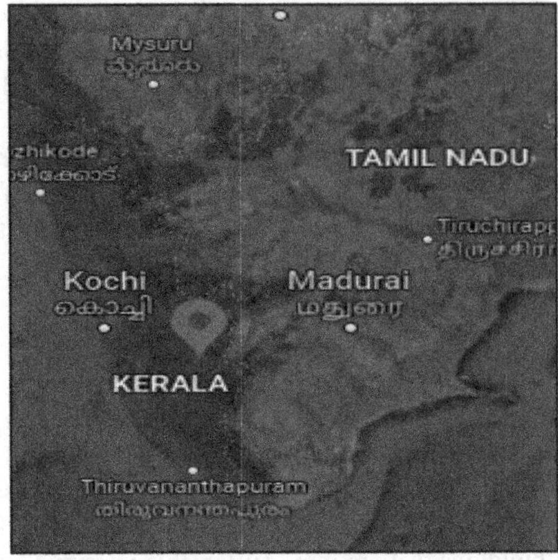

FIGURE 4.1

Location of the site.

9.6354°N and 76.9789°E (Latitude and Longitude). Climate analysis of a particular region can help to determine the input design parameters for energy analysis. This information helps the designer to interpret the effects of the weather and to assess the outputs of energy analysis.

4.2.1 Sustainable infrastructure

Good health and well-being, affordable and clean energy, and sustainable cities and communities are some of the sustainable development goals defined by the United Nations. To achieve this goal, it is important to create tools and methods for effective project management in the construction sector. BIM is a process that involves the generation and management of digital representations of physical and functional characteristics of places, which in turn aid AEC (architect, engineer, and construction) professionals to plan, construct, and manage buildings and infrastructure more efficiently. Safety, serviceability, and sustainability are the key aspects to be ensured in the life cycle of the infrastructure. It is now possible to achieve the said parameters and goals by integrating the data in the BIM process after appropriate analysis for effective management.

4.2.2 Building information modeling

The performance of the buildings under different environmental conditions is greatly influenced by the emission of greenhouse gasses, energy efficiency of the buildings, materials used for construction, water, and land use. The overall performance of the buildings mainly depends on their energy efficiency. The BIM model integrates all the tools to analyze the performance of buildings with the available data in one platform. It helps the AEC professionals to design the building effectively. The concept of detailed design using the BIM model provides the platform for effective construction management. The key performance indicators may be helpful in assessing the performance level of buildings as per the standards. Sun and shadow analysis, simulation of daylight, and glare for different orientations of the building are used to examine the energy efficiency of the BIM model. Fig. 4.2 depicts the visual representation of the BIM concept.

4.3 Methodology

4.3.1 Architectural model

An architectural model has been developed to simulate the physical representation of the residential building as shown in Fig. 4.3. It is the scaled model. It is modeled to study the aspects of architectural design and to impart design ideas. The developed 3D model is the digital representation of the building area in the site, including its terrain features. The model consists of information related to the geometry of the functional area, size of elements, and specifications of doors and windows. Revit Architecture modeling provides important data for dealing with a whole construction project in one platform. As a result of the broad nature of a 3D model, AEC professionals can use Revit as a framework to control communication with various individuals involved in the projects.

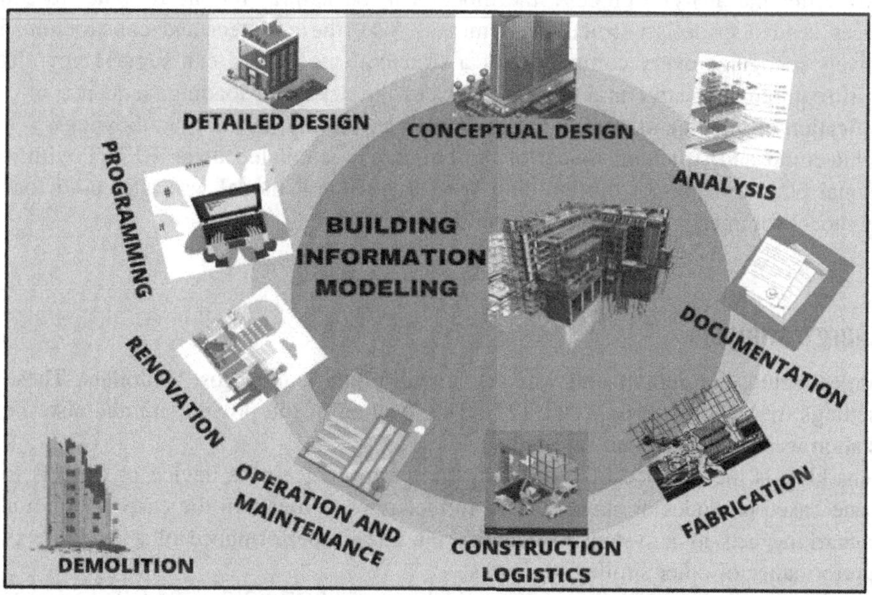

FIGURE 4.2

Visual representation of BIM concept.

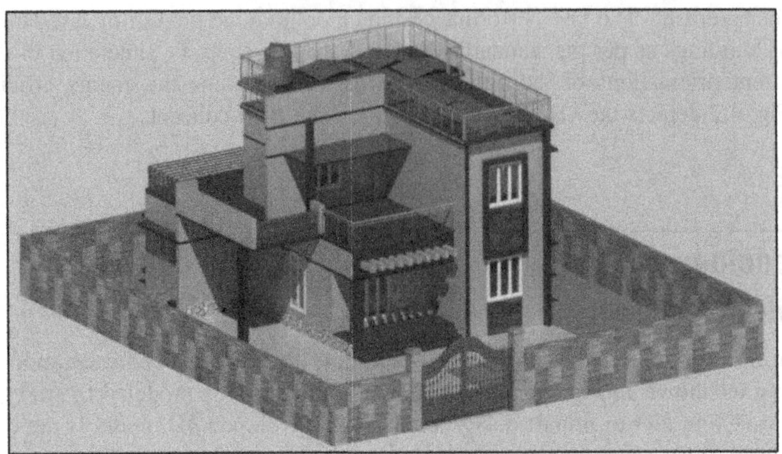

FIGURE 4.3

3D view of the building.

4.3.2 Structural model

As shown in Fig. 4.4 a structural model is developed with the sizes and thicknesses of structural members. It is a reinforced concrete type framing system consisting of slabs, beams, columns, and foundations. After the analysis process, the model will be helpful for the designer to extract the design forces needed to design structural members. Also, the architect and construction manager can effectively recognize every component of a structural model and can suggest any changes in the model if required. The structural system produces the details of loading, sequence of construction, specification of ingredients, and the relevant changes, if any, in the final drawings.

An architectural and structural model of the building was created using REVIT with reference to the original plan drawings of the building. The architectural model was later used for creating the energy model for energy analysis of the building.

4.3.3 Energy analysis

For the project setup, the default unit was set as metric and annual cost in dollars. These are the primary settings in Autodesk Insight 360 [27]. The location of the project site and also the nearest weather station were identified and selected.

Benchmarking is the practice of contrasting the performance of a device or process to inform and persuade based on codes or standards for further improvement. In the case of building energy use, benchmarking acts as a system to quantify the energy performance of a structure compared with the performance of other similar structures.

Benchmark value shall be kept as per the guidelines of ASHRAE Standard. There are two parameters considered for assessing the performance of a material: The *R*-value, which is the thermal resistance (ability to prevent the transfer of heat) of a material, and the *U*-value, which is the

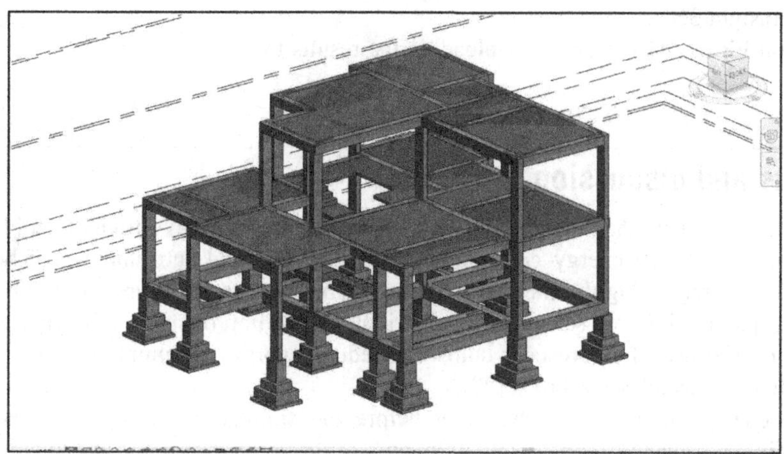

FIGURE 4.4

Structural model of the building.

thermal transmittance (ability to transfer heat) of a material. The SI unit of both properties is m^2K/W. All the specifications were obtained from the test results of the manufacturers.

Sustainability has become more important in infrastructure development in recent years. It is reported that about 30%–40% of all energy worldwide is used for building construction. Considering these aspects, a sustainable design approach is adopted to reduce the energy demand in buildings. The main principle in the design of the green building is to reduce the energy consumption of the structure by efficient design and on-site renewable energy generation.

The use of BIM in the construction field has gained a lot of attraction in recent years. This study focuses on designing a residential building using BIM tools to make it energy efficient. Various design parameters have been incorporated to develop the energy model. The models are analyzed using Autodesk Insight 360 to measure their energy efficiency. An attempt has been made to develop the scheme to efficiently manage the water cycle in a residential unit system. A BIM-based water efficiency framework has been developed to optimize traditional water efficiency based on an information database. This approach delivers an effective design for a zero-energy system in buildings to satisfy the goals of sustainable development.

In the proposed work, the energy model is developed using a computational framework, which consists of network captures to process heat transfer analysis throughout the building. The workflow comprises creating energy models, performing energy analysis, and optimizing building energy. Building performance analysis (BPA) tools are employed for performing the simulation using Autodesk REVIT. The basic simulation setup or workflow of the process is given below:

- Creating a model using combined massing and building elements
- Computational analysis → energy settings, setting → location, review → energy analytical model settings
- Creating the energy model; creating views → 3D energy model, spaces, and surfaces
- Reviewing the energy model

- Launch →Insight 360
- Commitment Firms (AIA 2030) → Uploading the results to DDx

4.4 Results and discussion

It is a prime requirement that the design and construction of energy-efficient buildings should include measures to reduce energy consumption. The location and surroundings of buildings are important factors in regulating the thermal exposure of building temperature and illumination. The use of low-cost passive infrared sensors may contribute to the improvement of energy efficiency in building design. The use of fluorescent bulbs can reduce energy consumption in the building by two-thirds of that of incandescent bulbs [28].

The space energy characteristics were set before the simulation study as per the standards, including lighting, occupancy, gains, ventilation, and infiltration parameters. The lighting properties have the following set data: (1) illumination as 5.5 W/m^2 per area, (2) end-usage category as lightly conditioned, (3) allowance with added decorative lighting, (4) luminaire category as recessed fluorescent lighting. The return air fraction is assumed to be zero, and the radiant-visible-replaceable fraction is considered 0.38, 0.17, and 0.46, respectively. The specifications imply that no illumination heat is transported to any neighboring rooms. A family of six people is considered for the modeling. It includes the number of people and activity types as sitting at rest, walking, light, medium, or heavy activity added, depending on what is expected from the inhabitants.

The gain attributes set for the design-level calculation method and fraction latent/radiant/lost were 0.4, 0.6, and 0.4, respectively. The ventilation and infiltration parameters are calculated using a design flow rate calculation technique that assumes a flow rate per person of 1.2 m^3/s, a ventilation type of natural ventilation, a fan pressure increase of 450 Pa with an overall efficiency of 0.5, and an airflow rate per room of 0.00015 m^3/s was considered. During the simulation, only the elements under consideration were changed, while all other parameters were kept constant. The investigation was conducted in the month of April, considering it is the hottest month of the year in Kerala. Simulating the energy study of the initial model reveals the peak energy requirement of the building as 29.8 USD/m^2/year for 624 kWh/m^2/year. As per ASHRAE Standard 90.1, the benchmark value for the energy-efficient building is 14.1 USD/m^2/year for consuming 328 kWh/m^2/year, as shown in Fig. 4.5.

4.4.1 Window-to-wall ratio

To stabilize the simulation process, the influence of the window size variation of the building under investigation was explored. Based on the situations, various WWRs were analyzed after different trials it was clear that as the WWR increases, energy efficiency decreases, ultimately increasing the overall cost. WWR can be adjusted for all four cardinal directions, and redistribution of the glazing area resulted in the reduction of the building's monthly peak energy consumption. It is reported that WWR has a great influence on the thermal comfort of buildings by 20%–55% [29]. The variation in the modified energy requirements based on the design modifications made to the baseline model is presented in Table 4.1.

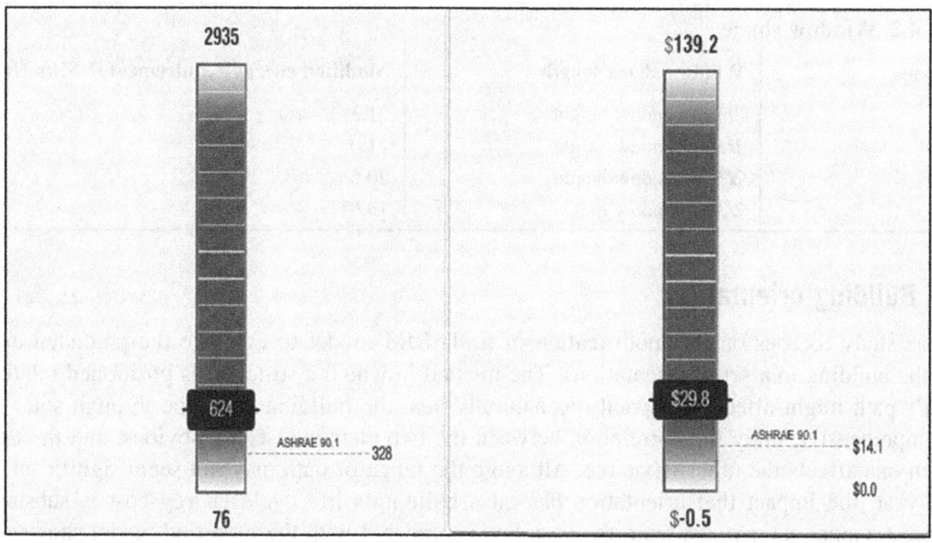

FIGURE 4.5

Initial benchmark.

Table 4.1 Window-to-wall ratio.

Directions	Design modifications (%)	Modified energy requirement (USD/m^2/year)
South	13	29.2
North	17	27.9
West	13	24.4
East	15	21.3

4.4.2 Window shade

Outdoor solar shades are perfect for any outdoor location as they eliminate much of the heat and glare that come with abundant sunlight. Exterior shades conveniently shield limited outdoor spaces as well as surrounding windows, doors, and walls from harmful ultraviolet radiation. Window shades in the base model were changed in all four directions, which considerably decreased energy usage. The influence on energy consumption also depends on other parameters, such as window size and solar heat gaining capabilities. Window shades can lower the total heat gain during the summer substantially on south-side windows and west-bound windows. Tan et al. [30] analyzed the impact of the physical characteristics of window shades on the energy consumption of residential buildings. It is proved that window shades can effectively reduce the energy consumption of air conditioners. Table 4.2 depicts the adjusted energy needed depending on the change in window shade length in the base model of the structure.

Table 4.2 Window shade.

Directions	Window shade length	Modified energy requirement (USD/m^2/year)
South	2/3 of window height	21.3
North	2/3 of window height	21.0
West	2/3 of window height	20.6
East	2/3 of window height	19.9

4.4.3 Building orientation

The case study focuses on the modification of a 3D-BIM model to evaluate the predicted energy use of the building in a set of orientations. The method in which a structure is positioned relative to the sun's path might affect its capacity to naturally heat the building envelope through solar gain. In a comprehensive study the correlation between the two elements seems obvious, and the degree to which one affects the other is visible. Although the range of data may not seem significant from year-to-year, the impact that orientation has on a building's life cycle energy cost is substantial. The lowest annual energy cost from the base level compared with the modified model shows a significant reduction in energy use. Higher efficient designs were obtained for varying the orientation of the residential building in the study conducted by [31]. In the present investigation, the optimum orientation of the building is south facing considering the modified energy requirement of 18.5USD/m^2/year as shown in Fig. 4.6.

4.4.4 Window glass

The base model of the windows is double-glazed glass in standard wood frames. The study compares the materials with triple low-emissivity (low-E) glass, double-glazed glass in aluminum frames, and double-glazed glass in timber frames. Table 4.3 shows the modified energy consumption influenced by window components and properties. The presence of low-E glass can greatly conserve heat with better solar control [32]. The presence of windows improves the building's ventilation and enables light to enter.

4.4.5 Lighting efficiency, daylight, and occupancy controls

Lighting efficiency represents the average internal heat gain and power consumption of electric lighting per unit area of the floor. Therefore, it plays a key role in energy-saving opportunities. Due to their higher energy consumption, incandescent bulbs have to be substituted with energy-efficient bulbs in the modified model. The design of indoor lighting for energy efficiency is based on some basic design principles and methods, such as more light is not necessarily better, the quality of light is as important as quantity, and maximizing the use of daylighting. Pilechiha et al. [33] developed a framework for optimizing the window size based on daylight. The improved energy consumption was 3.23 W/m^2 with a modified energy requirement of 14.4 USD/m^2/year. Fig. 4.7 represents lighting efficiency.

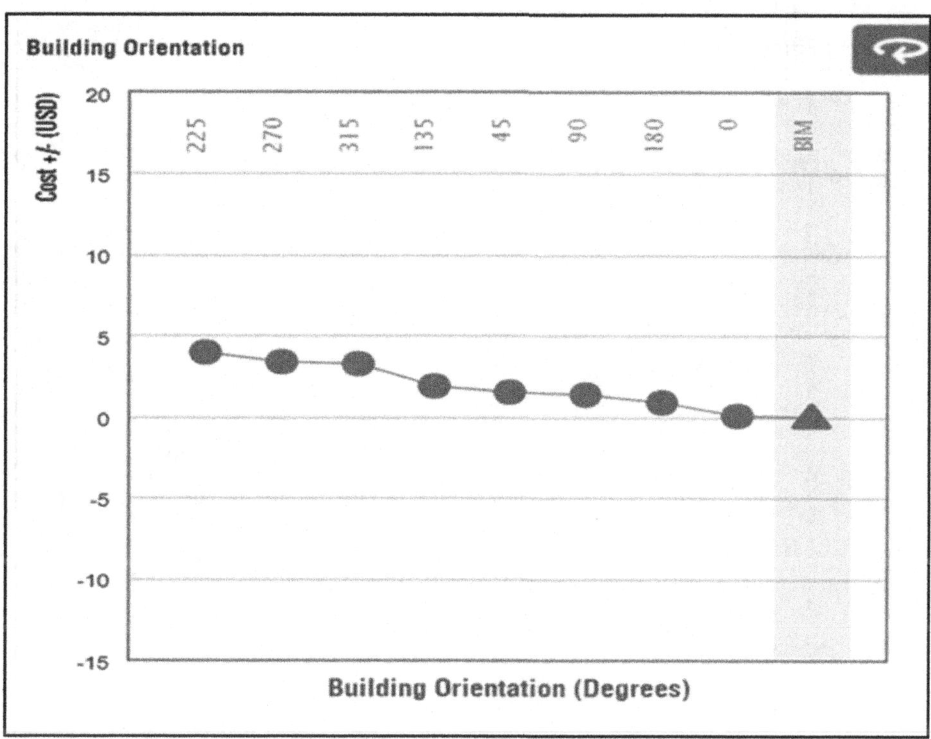

FIGURE 4.6

Building orientation.

Table 4.3 Window glass.

Directions	Glass type	Modified energy requirement (USD/m^2/year)
South	Triple low-emissivity	18.3
North	Triple low-emissivity	17.8
West	Triple low-emissivity	17.1
East	Triple low-emissivity	16.5

Daylight dimming and occupancy sensor systems were considered to control daylight and occupancy patterns. All control measures decrease the overall energy utilization through a blend of lighting energy savings through occupancy-based switching and daylight dimming. Energy savings in lighting control systems may be realized by adapting two dynamic aspects of the environment—varying amounts of daylight and changes in occupancy. Proper control using appropriate controlling devices can significantly reduce the energy consumption of lighting systems. The energy requirement was reduced to 14 USD/m^2/year due to the provision of both daylight and occupancy controls in the modified model. Fig. 4.8 represents daylighting and occupancy controls.

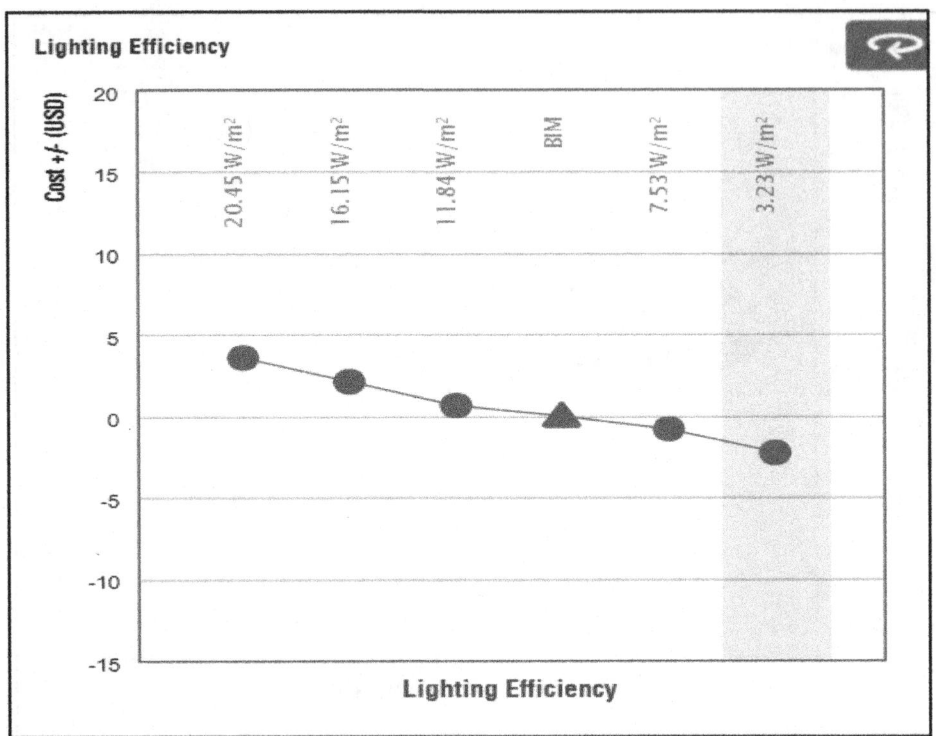

FIGURE 4.7

Lighting efficiency.

4.4.6 Wall and roof materials

In the research, several types of walls are compared, as illustrated in Fig. 4.9, and the findings reveal that walls with aerated blocks and concrete roofs deliver more energy efficiency. The adjusted energy need is greatly decreased to 12.6 USD/m^2/year.

Fig. 4.10 illustrates the change from the baseline model to the improved energy-efficient model. The most efficient combination of all the material choices was assessed in the simulation, and the final energy need after design adjustments was determined as 12.6 USD/m^2/year with a modified energy demand of 317 kWh/m^2/year. The conclusion suggests that the materials of the evaluated components for this building may be increased, thus reducing its energy consumption. One of the keys to addressing challenges in 3D modeling and energy analysis is the utilization of software tools with varying complexity, substantial uniformity, and ease of integration. In this study, integration with the construction cost would be useful since each design and material option is connected with a construction cost variance. So, a building cost estimate linked to the life cycle cost provides a complete perspective for the owners and investors. Using alternate materials or methods might increase the building cost and have the opposite effect on the project.

Further studies need to be done to compare the energy model findings to the actual consumption after the buildings are completed to enhance or refine this link, make it more realistic, and allow

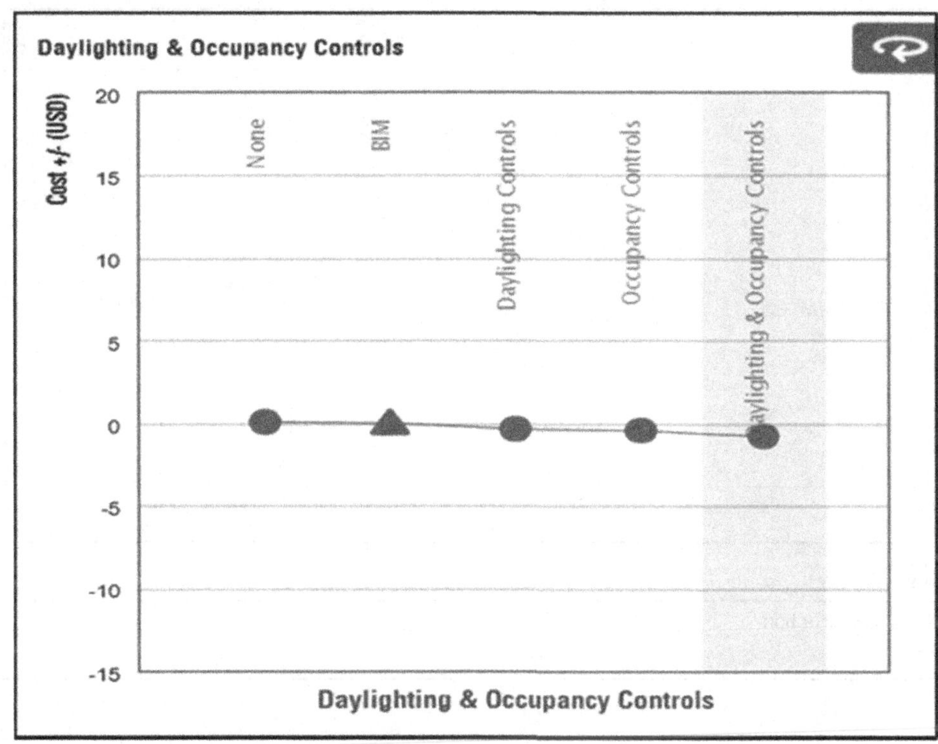

FIGURE 4.8

Daylighting and occupancy controls.

better judgments throughout the design process. These energy measurements may be prolonged over a lengthy period to make sure the life cycle energy consumption is still correct. The energy efficiency process depends on the construction type and location. It is generated to discover the aspect of a structure, and it has the greatest influence on decreasing energy consumption. Extra aspects or portions of a structure are included in the study to increase the variables; heating, ventilation, and air-conditioning (HVAC) systems and renewable systems like solar panels and structural elements built from wood are also integrated into the study. The final benchmark values satisfying the ASHRAE standards are obtained as USD 12.6 as shown in Fig. 4.11.

4.4.7 Sunlight exposure study

The sunlight exposure analysis visualizes and quantifies the distribution and intensity of solar radiation on building model surfaces. The study analyzes shadow cast by nearby objects such as vegetation and surrounding buildings. Even though the solar analysis is not designed for sizing photovoltaic panels, it may assist in finding areas for optimal solar gain by considering shading effects and seasonal fluctuations in solar radiation. As much as 90% of a building's environmental

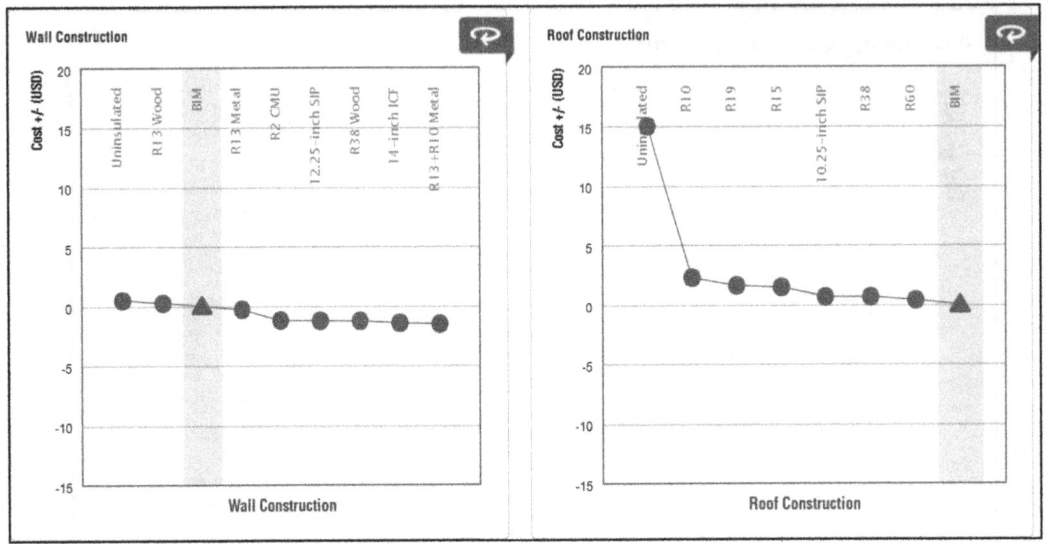

FIGURE 4.9

Wall and roof construction.

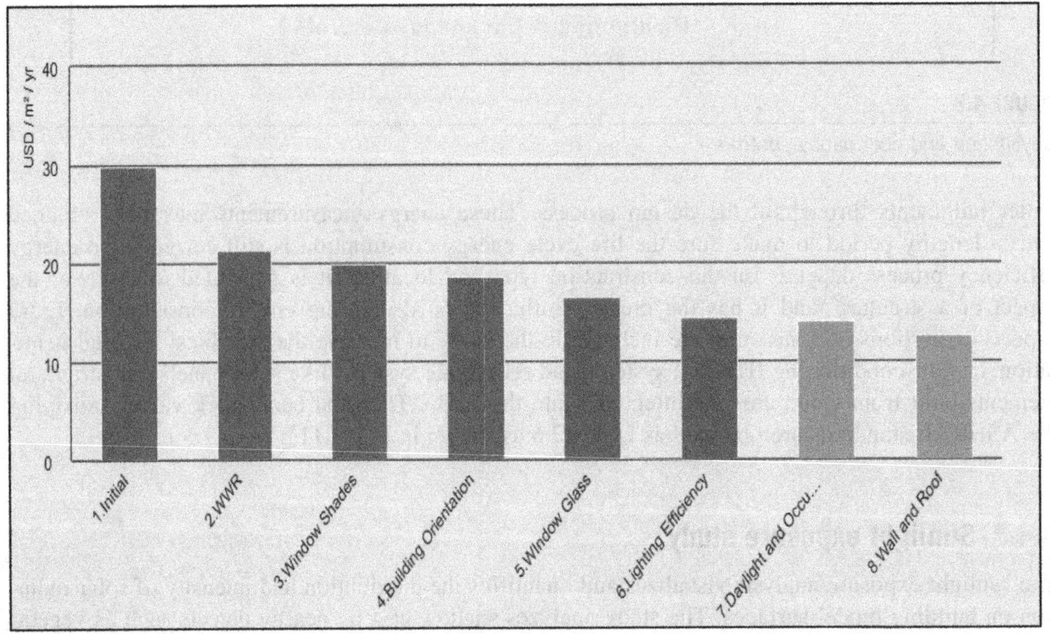

FIGURE 4.10

Scenario comparison.

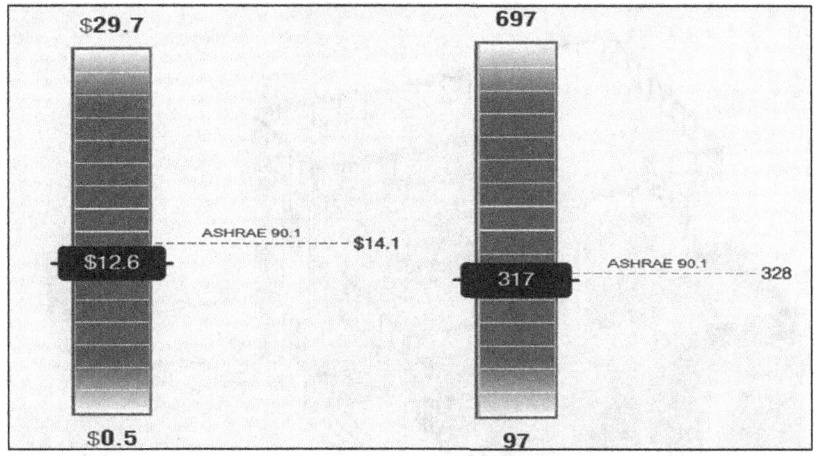

FIGURE 4.11

Final benchmark.

effect comes from the energy it utilizes over its lifespan. A passive design employs energy that is already accessible, such as sunlight and heat. The proper use of passive techniques may reduce or even eliminate the cost of heating or air conditioning and save money. If the sun strikes the building via its outside shell and windows, smart passive design may potentially transform the structure into an energy asset. Fig. 4.12 represents the model generated from the sunlight exposure study.

4.4.8 Bi-Directional plumbing system

The present investigation attempted to design a plumbing system where the greywater is recycled, and the same water is utilized again in the residence, as shown in Fig. 4.13.

Water from the storage tank goes to the various inlets. The black water from the toilet is transferred to a septic tank. The greywater from other outlets, including bathroom drains, sinks, and washbasins, are collected and transported to a small-scale treatment facility. This greywater is recycled and pumped back to the storage tank. From there, the flow continues as before. Figs. 4.14 and 4.15 represent the side view and back view of the conceptual plumbing model.

4.4.9 IOT-BIM integration system

The construction industry is experiencing a continuous digital transformation with the evolving IoT and BIM devices bringing together real-time information to assist the industry. IoT sensors and BIM can achieve real-time data integration and communication with various aspects. Advanced sensor technologies and BIM provide complete automation in the construction industry; an IoT device like RFID is an effective tool for prefabrication.

The key to success in the construction industry is a collaborative system to build the infrastructure with IoT-BIM, involving various technologies such as cloud computing, sensor integration, interoperability

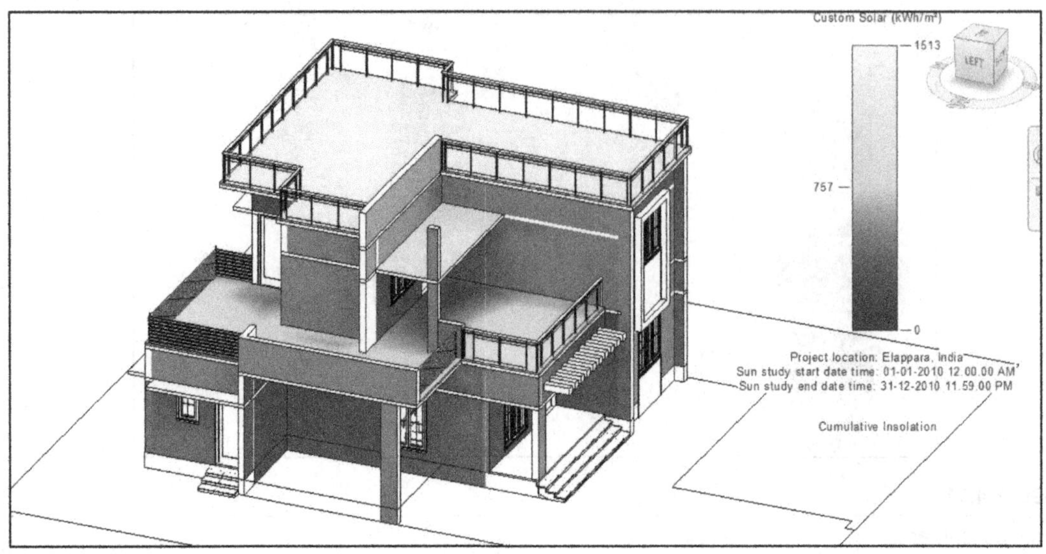

FIGURE 4.12

Sunlight exposure study.

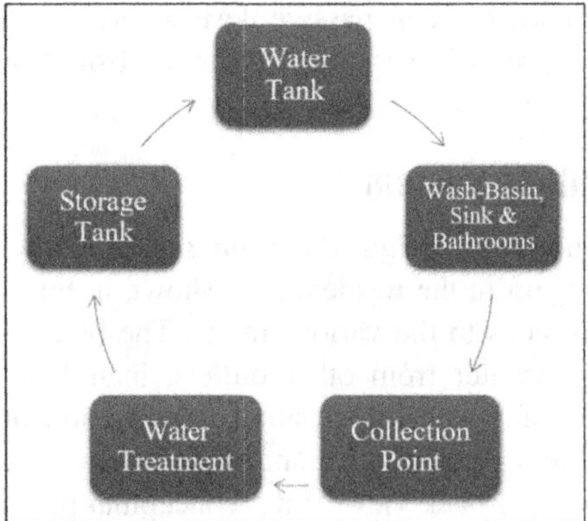

FIGURE 4.13

Water flow cycle.

issues, facility management, web analytics, energy control, and sustainability management. The prime technology for the integration of IoT-BIM is cloud-based analytics, which helps to integrate the building models and IoT sensors. The integration begins with the placing of sensors, equipment installation, and then moving into operation. To build a comprehensive BIM-based building lifecycle approach is to manage IoT for the building industry and to connect with cloud-based management systems. Fig. 4.16 shows the detailed schema for IOT-BIM integration systems.

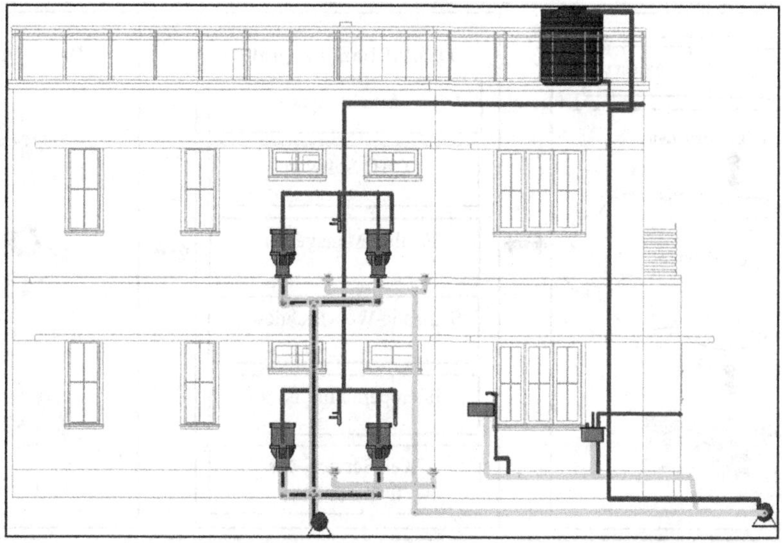

FIGURE 4.14

Side view of the conceptual plumbing model.

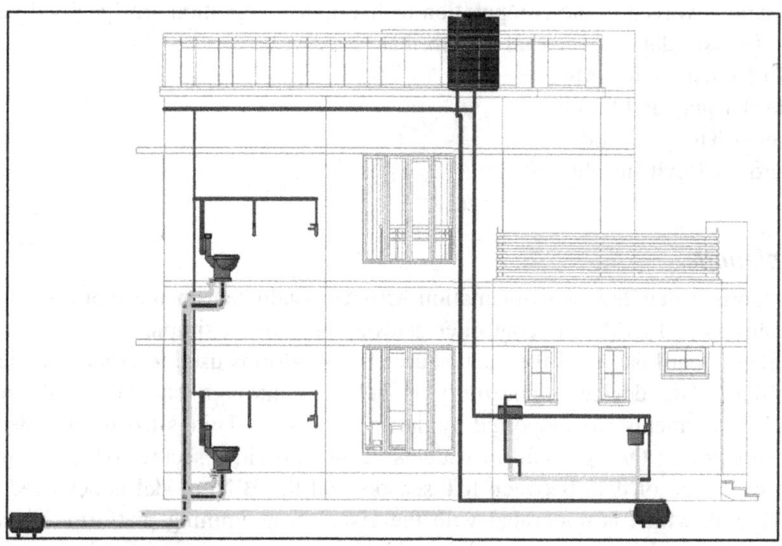

FIGURE 4.15

Back view of the plumbing model.

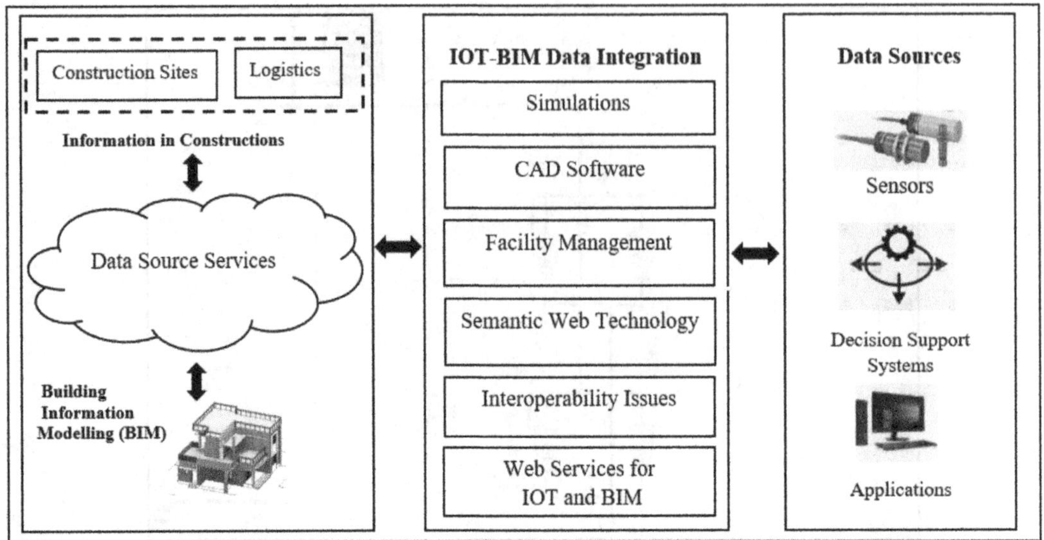

FIGURE 4.16

Schema for IoT BIM-based integration system.

Fig. 4.16 highlights the software tool for IoT-BIM integration systems. Web services, facility management, CAD software, and simulation units.

The major steps involved in the computation of IoT-BIM integration are specified below:

Collection of sensor data → Internal API format

Internal API format → IoT Module

IoT Module (Display and Storage)

IoT Module → Things Board

Things Board → Revit Insight 360

4.4.9.1 IoT platform

This chapter reports a detailed experimentation activity conducted on the cloud platform to integrate the IoT sensors and BIM. The controller gateway provides a simple configuration to monitor the sensors. Things Board is one of the open-source IoT platforms used to process, collect, manage, and visualize data. The device connectivity is achieved through standard IoT protocols; the cloud and local deployments are supported by the IoT platform. Things Board provides fault tolerance and data monitoring through built-in widgets. It also provides secure IoT entity provisioning. The automatic exchange of data between IoT sensors and the BIM model is achieved through the Revit INSIGHT 360, which is integrated with the visual programming platform. Visual programming is not involved with coding but with graphics-based entities. Each node is connected with input and output to perform specific tasks. Fig. 4.17 highlights the position of IoT sensors in the model. Fig. 4.18 shows the complete physical setup for the proposed method.

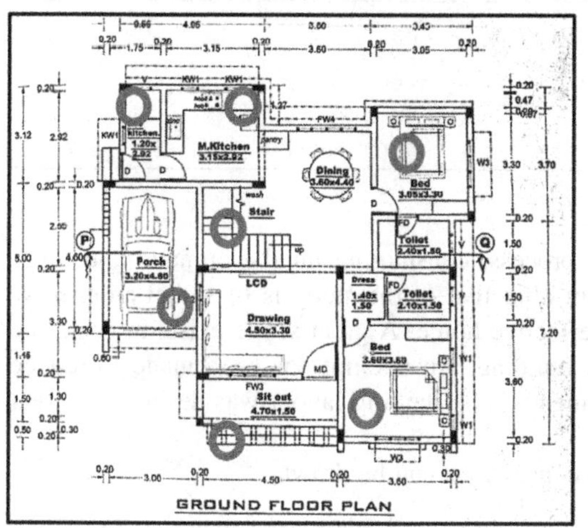

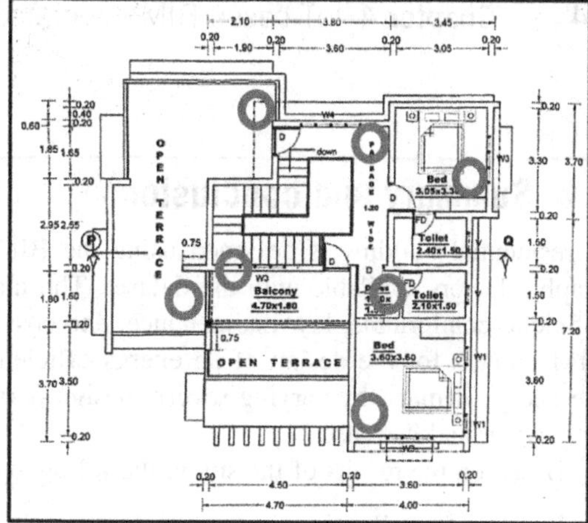

FIGURE 4.17

Placement of IoT sensors.

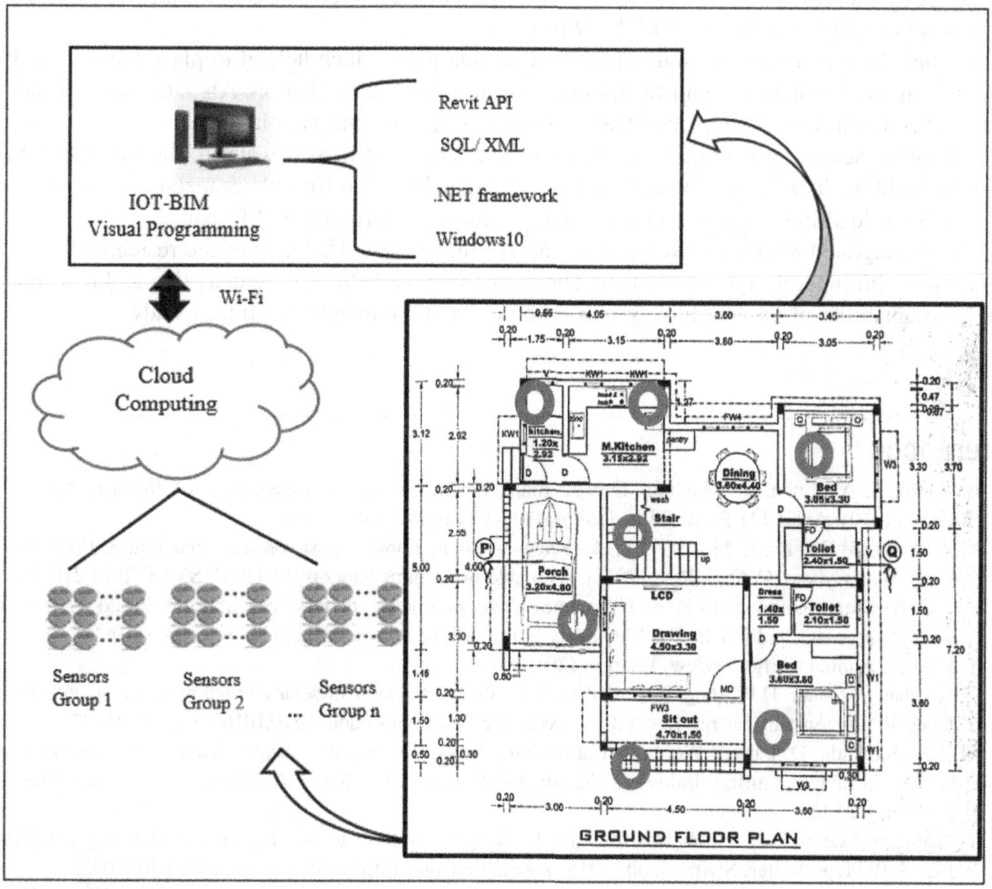

FIGURE 4.18

Physical setup for IoT-BIM integration method.

4.5 Summary and conclusions

A residential building is designed using the BIM process, considering all the climatic and topographic factors available with the dataset. The main objective of the study is to model an energy-efficient, comfortable, low-maintenance, and cost-effective home. A great scope exists in using the BIM process for the design of an energy-efficient building. An attempt has been made to reduce the energy demand by varying several design parameters. A plumbing layout was designed to treat gray water and its reuse.

Based on the results of the study, the following conclusions can be drawn:

- Window-to-wall ratio, window shades, building orientation, wall, glass, roof material, and daylight exposure are the key influencing factors that affect the energy efficiency of the building.
- The initial estimated cost in terms of energy consumption for the building was 29.8 USD/m^2/year. After the design modifications, the cost was reduced to 12.6 USD/m^2/year, which is below the limit of ASHRAE, that is, 14.1 USD/m^2/year.
- The sunlight occurrence was found based on the sun path, which helped to place the solar units in a position. It is helpful to allow optimal utilization of solar radiation. Also, the number and positions of windows were placed based on natural lighting and ventilation.
- A plumbing scheme was designed as a layout to connect the treated water to the water system of the building. The reuse of water reduces the dependency on freshwater resources and also reduces the load in the septic tank and soak pit, further improving its lifespan.
- A BIM-integrated model is developed to analyze the energy efficiency of the residential building with input design parameters. The model may be helpful in reanalyzing and redesign the sustainability of the building system with the input parameters for future study.

References

[1] B. Lashkari, Y. Chen, P. Musilek, Energy management for smart homes-state of the art, Appl. Sci. (Switz.) (2019). Available from: https://doi.org/10.3390/app9173459.

[2] M.R. Alam, M.B.I. Reaz, M.A.M. Ali, A review of smart homes—past, present, and future, IEEE Trans. Systems Man. Cybern. C Appl. Rev. (2012). Available from: https://doi.org/10.1109/TSMCC.2012.2189204.

[3] R. Lutolf, Smart home concept and the integration of energy meters into a home based system, in: Metering Apparatus and Tariffs for Electricity Supply. Seventh International Conference, 1992.

[4] T. Energy, Annual energy review, Energy. (2010).

[5] K.R. Ullah, R. Saidur, H.W. Ping, R.K. Akikur, N.H. Shuvo, A review of solar thermal refrigeration and cooling methods, Renew. Sustain. Energy Rev. (2013). Available from: https://doi.org/10.1016/j.rser.2013.03.024.

[6] M. Sachidananda, D. Patrick Webb, S. Rahimifard, A concept of water usage efficiency to support water reduction in manufacturing industry, Sustainability (Switz.) (2016). Available from: https://doi.org/10.3390/su8121222.

[7] T. Boyle, D. Giurco, P. Mukheibir, A. Liu, C. Moy, S. White, et al., Intelligent Metering for Urban Water: A Review, Water, Switzerland, 2013. Available from: https://doi.org/10.3390/w5031052.

[8] J.M. Baptista, The regulation of water and waste services: an integrated approach (Rita-Ersar), Water Intell. Online (2014). Available from: https://doi.org/10.2166/9781780406534.

[9] L. Luo, D. Apps, S. Arcand, H. Xu, M. Pan, M. Hoerling, Contribution of temperature and precipitation anomalies to the California drought during 2012–2015, Geophys. Res. Lett. (2017). Available from: https://doi.org/10.1002/2016GL072027.

[10] E.D. Maer, A.A. Pop, D.C. Popa, I.C. Gros, Hybrid water collecting and management system using smart home technologies, in: 2021 28th International Workshop on Electric Drives: Improving Reliability of Electric Drives, IWED 2021 - Proceedings, 2021. https://doi.org/10.1109/IWED52055.2021.9376351.

[11] L. Agel, M. Barlow, F. Colby, H. Binder, J.L. Catto, A. Hoell, et al., Dynamical analysis of extreme precipitation in the US northeast based on large-scale meteorological patterns, Clim. Dyn. (2019). Available from: https://doi.org/10.1007/s00382-018-4223-2.

[12] F. Cannon, L.M.V. Carvalho, C. Jones, A. Hoell, J. Norris, G.N. Kiladis, et al., The influence of tropical forcing on extreme winter precipitation in the western Himalaya, Clim. Dyn. (2017). Available from: https://doi.org/10.1007/s00382-016-3137-0.

[13] D. Koo, K. Piratla, C.J. Matthews, Towards sustainable water supply: schematic development of big data collection using internet of things (IoT), Proc. Eng. (2015). Available from: https://doi.org/10.1016/j.proeng.2015.08.465.

[14] T. Robles, R. Alcarria, D. Martín, M. Navarro, R. Calero, S. Iglesias, et al., An iot based reference architecture for smart water management processes, J. Wirel. Mob. Networks Ubiquitous Comput. Depend. Appl. (2015). Available from: https://doi.org/10.22667/JOWUA.2015.03.31.004.

[15] T. Robles, R. Alcarria, D. Martin, A. Morales, M. Navarro, R. Calero, et al., An internet of things-based model for smart water management, in: Proceedings - 2014 IEEE 28th International Conference on Advanced Information Networking and Applications Workshops, IEEE WAINA 2014, 2014. https://doi.org/10.1109/WAINA.2014.129.

[16] H.M. Ramos, A. McNabola, P.A. López-Jiménez, M. Pérez-Sánchez, Smart water management towards future water sustainable networks, Water (Switz.) (2020). Available from: https://doi.org/10.3390/w12010058.

[17] X. Xu, T. Mumford, P.X.W. Zou, Life-cycle building information modelling (BIM) engaged framework for improving building energy performance, Energy Build. (2021). Available from: https://doi.org/10.1016/j.enbuild.2020.110496.

[18] C. Wang, Y.K. Cho, C. Kim, Automatic BIM component extraction from point clouds of existing buildings for sustainability applications, Autom. Constr. (2015). Available from: https://doi.org/10.1016/j.autcon.2015.04.001.

[19] J. Gubbi, R. Buyya, S. Marusic, M. Palaniswami, Internet of Things (IoT): a vision, architectural elements, and future directions, Fut. Gener. Comput. Syst. (2013). Available from: https://doi.org/10.1016/j.future.2013.01.010.

[20] C.K. Dehury, P.K. Sahoo, Design and implementation of a novel service management framework for IoT devices in cloud, J. Syst. Softw. (2016). Available from: https://doi.org/10.1016/j.jss.2016.06.059.

[21] A. Čolaković, M. Hadžialić, Internet of Things (IoT): a review of enabling technologies, challenges, and open research issues, Comput. Netw. (2018). Available from: https://doi.org/10.1016/j.comnet.2018.07.017.

[22] S. Tang, D.R. Shelden, C.M. Eastman, P. Pishdad-Bozorgi, X. Gao, A review of building information modeling (BIM) and the internet of things (IoT) devices integration: present status and future trends, in: Automation in Construction, 2019. https://doi.org/10.1016/j.autcon.2019.01.020.

[23] J. Teizer, M. Wolf, O. Golovina, M. Perschewski, M. Propach, M. Neges, et al., Internet of Things (IoT) for integrating environmental and localization data in Building Information Modeling (BIM), in: ISARC 2017 - Proceedings of the 34th International Symposium on Automation and Robotics in Construction, 2017. https://doi.org/10.22260/isarc2017/0084.

[24] IGBC, Indian green building council, J. Chem. Inf. Modeling (2013).

[25] American society of heating, refrigerating and air-conditioning engineers, International Journal of Refrigeration, 1979. https://doi.org/10.1016/0140-7007(79)90114-2.

[26] BIS, National Building Code of India, 2016 Volume 1, National Building Code of India, 2016.

[27] A. Revit, Autodesk revit architecture 2010, Architecture. (2010).

[28] A.G. Abo-khalil, S.S. Ahmed, A new approach to improve the energy efficiency of middle-east buildings, The 7th Conference on Future of Renewable and New Energy in the Arab World, Assiut University, Assiut, Egypt, February 2013, 1–16, 2014.

[29] S. Pathirana, A. Rodrigo, R. Halwatura, Effect of building shape, orientation, window to wall ratios and zones on energy efficiency and thermal comfort of naturally ventilated houses in tropical climate, Int. J. Energy Environ. Eng. (2019). Available from: https://doi.org/10.1007/s40095-018-0295-3.

[30] Y. Tan, J. Peng, C. Curcija, R. Yin, L. Deng, Y. Chen, Study on the impact of window shades' physical characteristics and opening modes on air conditioning energy consumption in China, Energy Built Environ. (2020). Available from: https://doi.org/10.1016/j.enbenv.2020.03.002.

[31] J. Morrissey, T. Moore, R.E. Horne, Affordable passive solar design in a temperate climate: an experiment in residential building orientation, Renew. Energy (2011). Available from: https://doi.org/10.1016/j.renene.2010.08.013.

[32] Y. Yaşar, S.M. Kalfa, The effects of window alternatives on energy efficiency and building economy in high-rise residential buildings in moderate to humid climates, Energy Convers. Manag. (2012). Available from: https://doi.org/10.1016/j.enconman.2012.05.023.

[33] P. Pilechiha, M. Mahdavinejad, F. Pour Rahimian, P. Carnemolla, S. Seyedzadeh, Multi-objective optimisation framework for designing office windows: quality of view, daylight and energy efficiency, Appl. Energy (2020). Available from: https://doi.org/10.1016/j.apenergy.2019.114356.

CHAPTER

5 Reliable data sharing in medical cyber physical system using fog computing

Rachana Y. Patil[1], Arijit Karati[2], Yogesh Patil[3] and Aparna Bannore[4]

[1]*Department of Computer Engineering, Pimri Chinchwad College of Engineering, Pune, Maharashtra, India*
[2]*Department of Computer Engineering, National Sun Yat-sen University, Sizihwan, Kaohsiung, Taiwan*
[3]*IMEI Department, VBK Infrastructures, Pune, Maharashtra, India* [4]*Department of Computer Engineering, SIES Graduate School of Technology, Mumbai, Maharashtra, India*

5.1 Introduction

According to a research by Fierce Healthcare [1], most of the time in the health care industry, insecure handling of patients' electronic health records (EHR), leads to dishonest access, which has nearly tripled in 2019 compared with prior years [2,3].

Furthermore, due to remote working in the COVID-19 situation, the violations are much more, where there are incidents of patients' personal data being marketed online for money and patients are extorted with threats of exposing their data in the community [4]. Another issue with conventional EHR systems is that healthcare providers control and store patients' health records which may slow down the availability of data from the EHR systems affecting patients from getting healthcare facilities on time [5]. Interoperability across multiple her systems is also a concern. Owing to the aforementioned issues, herEHR system with sophisticated security at its heart, along with the decentralization of patient data control, is required [6].

Most of the IoT applications are cloud based. Cloud computers deliver various kinds of services, including networking, storage, computing resources, and high-quality IoT services [7,8]. Though, there are certain constraints like as network overcrowding, reduced use of bandwidth, fault acceptance and security challenges, etc.

Many IoT-dependent applications, for example, healthcare monitoring applications, which involve real-time data assessment and analysis within a short time interval and the major difficulties due to communication with cloud during data transfer, implementing tasks and returning cloud reply to the request is intolerable. Cisco developed a new computing idea called fog computing to address cloud computing restrictions.

Fog computing is an important computing architecture that has contributed substantially to IoT healthcare and monitoring services. Healthcare services, such as household nursing services for the aged, and chronic cardiac disease patients, require latency sensitive and real-time surveillance as well as data analysis and decision-making [9,10].

Intelligent Edge Computing for Cyber Physical Applications. DOI: https://doi.org/10.1016/B978-0-323-99412-5.00007-1

In the proposed study we suggest a system that can address the security, confidentiality, interoperability, and capability requirements of an EHR system using proxy signcryption technique. Patient data can be shared with hospitals, research organizations, and government agencies when requested. The proposed scheme data owners delegate the rights of signcryption to fog nodes, which in turn send the data to cloud server for permanent storage.

5.2 Related works

In recent times the topic of safety and secrecy in EHRs have sparked a lot of debate in their acceptance. The issue of access to claims data, by what means data is store, its transfer security, analysis rights, and regulating regulations posed a continuous obstacle in electronic health acceptability and remain unanswered research concerns. Now a days such systems are necessary to expand to fit in today's EHR system standards for EHR to be safer and more generalized for use [11]. Several pilot projects and models have investigated securing health records storing and access.

The authors of [12,13] presented a security approach that leverages Role Based Access Control to alleviate nearly all of the privacy and security difficulties connected along with web-based electronic health care records. Instead of handing over control of the EHR to all health-care staff, privacy is accomplished by assigning responsibilities to individual person who may want access to data, with each role describing the set of benefits and processes that an individual presuming that role is allowed to perform. The complete system was designed around the concept of roles, authorization management, and role hierarchy and inheritance. The authors also agreed that cryptographic and authentication procedures are required to provide a secure manner of communication over an insecure public internet. On the other hand, their intended Role Based Access Control model just appears as a right of entry to control matrix that regulates which items a certain role can access, and does not take into account data subject privacy control [13]. Came up with plan that may effect in deadlock in real time application while attempting to realize security and privacy, since EHR privacy is also possible to manage collectively by both the data subject and the health care expert as per [9], who also hold up cryptographic method for security assessment. The authors of [14] proposed a methodology to protect health information system hypothesized access control requirement and debated that access control system are similar to the one defined in [15] cannot sufficiently consider real world techniques for roles due to the difficulty in defining constraints. The authors stated that regardless of the sensitivity of the data and the emerging threat, not much consideration has been given to the difficulties of real-world access constraints. The scheme described in [16] put up the concept to protect health information system based on access control obligation and disagreed with one that described in [17] that access control system cannot sufficiently think real world methods for roles due to the difficulty in describing constraints. The authors pointed that in spite of the sensitivity of the data and the increasing risk, not much consideration has been given to the difficulties of real-time access restrictions.

The security and privacy repercussion that may occur when incorporating new tools into the conventional health care system were exclusively recognized in [2]. The authors affirmed that the matter of data access, data storage and investigation are though not particular to the medical field only and that parallel problems have been critically measured in other scopes like the financial services and online shopping and that the technical solutions are available that can be useful to EHRs in order to tackle these concerns in multiuser settings [18].

The authors of [18] Verified patient's data secrecy and safety threats intrinsic in the transformation from paper health data storage to the electronic methodology. Once More, in contrast to the concept of [19], it was revealed that nothing is unrestricted, compulsory, or role-based entry management practices in segregation could efficiently fulfill the privacy and security [20] provisions of an EHR.

The proxy signcryption method (ID-PSC) proposed in [21] fails to meet the security requirement of unforgeability, as well as the concept of proxy safeguards. The another scheme of ID-BPS in line with the previous scheme is proposed by [22] which is computationally efficient. The authors of [23] proposed a scheme combining public verifiability and forward secrecy features of security. The authors of [24] proposed ID-BPS system delegated partial signing rights to a proxy agent. The signature produced by a proxy signer can be distinguished from the signature generated by the principle signer in the proposed approach. The scheme proposed by authors of [25] satisfies the security requirements of public verifiability and forward secrecy. The authors of [21] offer a method that involves less processing and communication cost. The improved scheme is proposed by [26] and termed a proven secure ID-BPS scheme under the random oracle model (ROM). In terms of processing power, the technique is effective.

Due to the usage of elliptic curve cryptosystem (ECC) and bilinear pairing, the aforementioned techniques can maintain high computation and communication speeds. This type of intense calculation may be too much for devices with low computational capacity. In addition, current schemes lack actual application situations such as IoT, IIoT, and so on. As a result of the aforementioned concerns, we suggested an Identity based proxy signcryption for MCPS using fog computing (IBPSC-MCPC-FC) system that is safe.

5.3 Complexity assumptions

1. *Hyper Elliptic Curve Discrete Logarithm Problem (HECDP) Assumption* [27–29]

   Consider there is a divisor D_h of order p from the group of Jacobian (f_p). Also, there is an equation $\mathcal{L} = \aleph * D_h$ where $\aleph \in f_p$, therefore computation of $\aleph$ is having a negligible advantage.
2. *Hyper Elliptic Curve Computational Diffie-Hellman (HEC-CDH) Assumption*

   Consider there is a divisor D_h of order p from the group of Jacobian (f_p). Also, there is an equation $\mathcal{L} = \aleph * \mathcal{M} * D_h$ where $\aleph$ and $\mathcal{M} \in f_p$, therefore computation of $\aleph$ and $\mathcal{M}$ is having a negligible advantage.

5.4 Formal model of proposed IBPSC-MCPC-FC scheme

The IBPSC-MCPC-FCScheme is divided into six algorithms as follows.

1. Setup Algorithm: This algorithm is accountable for generating public parameters which are openly accessible to all the participating entities and master secret M_{SK} which is a secret of the trusted third party.

2. Key Extraction Algorithm: Every individual user sends his/her unique identity ID_a to the trusted third party. The secret key for the user S_a is generated and returned via the secret channel.
3. Warrant Generation and Delegation Algorithm: The original signer shall make a message warrant M_w which contains the information about the type of delegation and time of delegation; it also defines the type of documents to be signcrypted by proxy signcryptor. This algorithm is accountable for generating the signing warrant S_w and delegating it to proxy signer.
4. Warrant Verification Algorithm: This algorithm is accountable for the verification of signing warrant received from original signer. If the warrant is verified correctly then the proxy signer executes the next algorithm.
5. Proxy Signcryption Algorithm: This algorithm takes the message to be sent m, proxy signers identity ID_{pr}, proxy Signers private key S_{pr} identity of receiver ID_u and public parameters as input and generates the signcrypted message and send to the receiver via a secure channel.
6. Unsigncryption Algorithm: This algorithm takes received signcrypted message, receivers private key S_u and the identity of both sender and receiver ID_o, ID_u and generates the original message m if the signcrypted message has not tampered else it returns $\perp$.

5.4.1 Security definition

The proposed IBPSC-MCPC-FC scheme must satisfy confidentiality and unforgeability of original message. Let us consider that there exist an adversary $\mathscr{A}_d$ for the proposed scheme and $\mathbb{C}_h$ is a challenger [30]. For indistinguishability against adaptive chosen cipher text attack (IND-CCA2) the following interaction between adversary $\mathscr{A}_d$ and challenger $\mathbb{C}_h$.

Definition 1: If adversary $\mathscr{A}_d$ with no polynomial time and having nonnegligible advantage win the following game, then the proposed scheme IBPSC-MCPC-FC can achieve the security requirements of IND-IBPSC-MCPC-FC -CCA2.

Initial: The challenger $\mathbb{C}_h$ executes the setup algorithm to get the public parameters and a master secret ϑ. Then $\mathbb{C}_h$ sends the public parameters to adversary $\mathscr{A}_d$ and keeps ϑ with itself.

Phase 1: Adversary $\mathscr{A}_d$ executes the following queries which are interdependent.

1. Key Extractionquery: Adversary $\mathscr{A}_d$ selects the unique identity as ID. The challenger $\mathbb{C}_h$ runs key extraction algorithm and returns the S_{ID} to Adversary $\mathscr{A}_d$.
2. Warrant Generation and Delegationquery: The adversary $\mathscr{A}_d$ sends the request for signing warrant. The challenger $\mathbb{C}_h$ returns the warrant w and signing warrant S_w.
3. Warrant Verificationquery: The adversary $\mathscr{A}_d$ verifies the signing warrant received from challenger $\mathbb{C}_h$.
4. Proxy Signcryptionquery: The adversary $\mathscr{A}_d$ selects message m and the identities ID_a, ID_b and ID_c. The challenger $\mathbb{C}_h$ executes Key Extraction and Warrant generation and Delegation to get secret keys of S_a, S_b and the signing warrant S_w, then executes Proxy Secret key Generation to get PSK_{OP}. Finally the challenger $\mathbb{C}_h$ runs Proxy Signcryption and sends the signcrypted ciphertext σ to $\mathscr{A}_d$.

5. Unsigncryptionquery: The adversary $\mathscr{A}_d$ selects the signcrypted ciphertext σ and the identities ID_a, ID_b and ID_c. The challenger $\mathbb{C}_h$ runs Key Extraction algorithm to get the S_c, then executes Unsigncryption algorithm and sends result to $\mathscr{A}_d$.

Challenge: The adversary $\mathscr{A}_d$ wishes to be challenged on the two messages M_0, M_1 and identities ID_i, ID_j. In the first stage $\mathscr{A}_d$ cannot query for secret key of any of the identity. The challenger $\mathbb{C}_h$ produces the random bit $b \in_R \{0,1\}$ for which the $\sigma = \text{signcrypt}(M_b, S_b, ID_C)$ and sends to $\mathscr{A}_d$.

Phase 2: The adversary $\mathscr{A}_d$ executes the queries like phase 1. Except Key Extraction query for identities ID_i, ID_j and unsigncrypted text for σ.

Guess: The adversary $\mathscr{A}_d$ produces he random bit $b' \in_R \{0,1\}$. If $b = b'$ the adversary $\mathscr{A}_d$ wins the game. We have following advantage of $\mathscr{A}_d$

$$\text{Adv}(\mathscr{A}_d) = \left|\Pr[b = b'] - \frac{1}{2}\right|$$

Definition 2: The proposed scheme IBPSC-MCPC-FC can achieve the existential unforgeability against adaptive chosen messages attack (EUF- IBPSC-MCPC-FC-CMA) if adversary $\mathscr{A}_d$ with no polynomial time and having nonnegligible advantage in the following game.

Initial: The challenger $\mathbb{C}_h$ executes the setup algorithm to get the public parameters and a master secret ϑ. Then $\mathbb{C}_h$ sends the public parameters to adversary $\mathscr{A}_d$. Then $\mathscr{A}_d$ performs polynomial limited number of queries like in IND-IBPSC-MCPC-FC -CCA2. Finally, adversary $\mathscr{A}_d$ generates (σ, ID_i, ID_j), In phase 2 the private key for ID_i was not asked and the adversary $\mathscr{A}_d$ wins the game if the output of Unsigncryption (σ, S_i, ID_j) is not $\perp$ (Table 5.1).

Table 5.1 Symbols used in proposed IBPSC-MCPC-FC scheme.

Symbol	Definition
ID_i	Identity of user i
P_o, S_o	The public key and secret key of data owner
P_u, S_u	The public key and secret key of data user
P_{pr}, S_{pr}	The public key and secret key of proxy
$\mathcal{H}_0$, $\mathcal{H}_1$, $\mathcal{H}_2$, $\mathcal{H}_3$, $\mathcal{H}_4$	One-way cryptographic hash function
Z_p	Set of integers $\{0, 1, \ldots p-1\}$
$\vartheta \in_R Z_p$	Element ϑ randomly selected from Z_p
M_{Pub}	Master Public key
D_h	Divisor of hyperelliptic curve
S_w	Signcrypting warrant
M_w	Message warrent
m	Plaintext message
σ	signcrypted ciphertext
$\oplus$	XOR operation
$\perp$	Error symbol

5.5 Proposed identity based proxy signcryption for MCPS using fog computing

The system model for the proposed Identity based Proxy Signcryption for MCPS using Fog Computing (IBPSC-MCPC-FC) is shown in Fig. 5.1. The detailed algorithm is discussed below

5.5.1 Setup algorithm

Input: Security parameters λ of hyperelliptic curve

Output: Public system parameters

1. Identity of each participating entities is represented as ID_i
2. Select $\vartheta \in_R Z_p$, where ϑ is a master secret key (M_{SK})

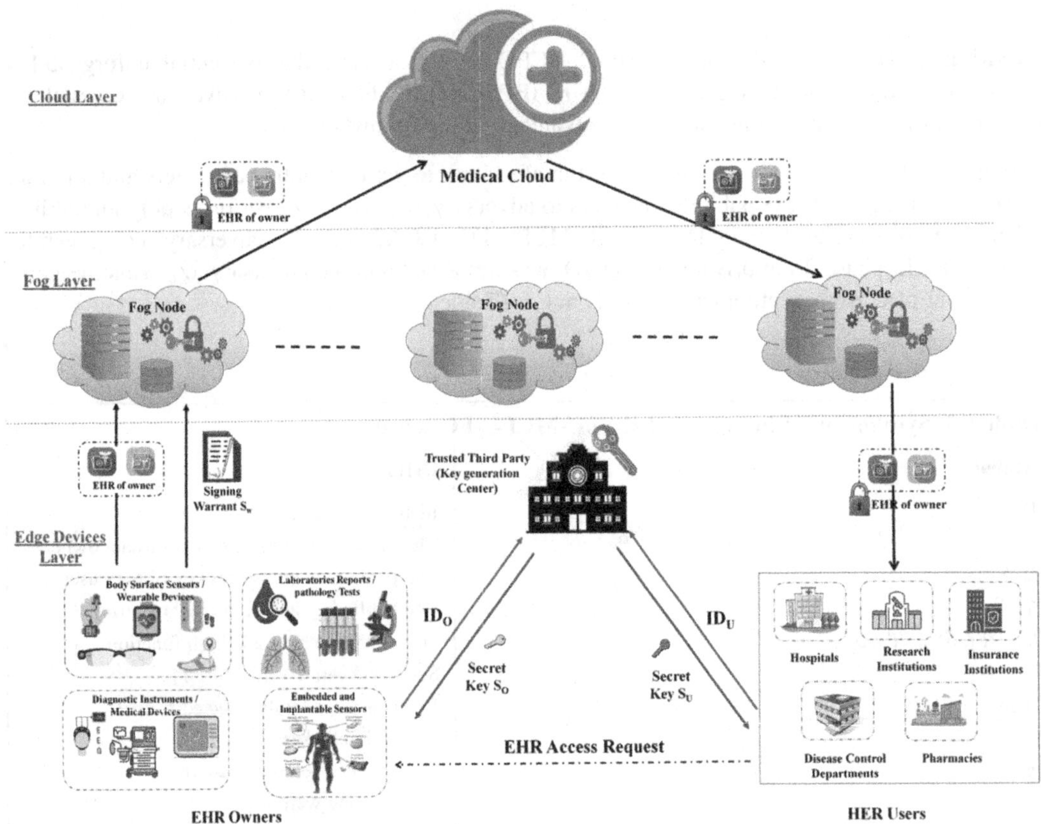

FIGURE 5.1

The system model for the proposed identity based proxy signcryption for MCPS using fog computing (IBPSC-MCPC-FC).

3. Master Public key $M_{\mathrm{Pub}} = \vartheta * D_{\mathrm{h}}$
4. $\mathcal{H}_0, \mathcal{H}_1, \mathcal{H}_2, \mathcal{H}_3, \mathcal{H}_4$ are secured one way hash functions (SHA-512)
5. The PKG publish the public system parameters as $\{M_{\mathrm{Pub}}, D_{\mathrm{h}}, \mathcal{H}_1, \mathcal{H}_2, \mathcal{H}_3, \mathcal{H}_4\}$, The ϑ is kept secret.

5.5.2 Key extraction algorithm

Input: Identity of participating entities ID_i

Output: Public and secret keys for ID_i

1. Compute $\alpha_o = \mathcal{H}_0(\mathrm{ID}_o, \vartheta)$
2. The Public key of user A with identity ID_o is $P_o = \alpha_o * D_{\mathrm{h}}$
3. Compute $\gamma_o = \mathcal{H}_0(\mathrm{ID}_o, P_o)$
4. The Secret key of user A with identity ID_o is $S_o = [\alpha_o + (\gamma_o * \vartheta)] \bmod p$

5.5.3 Warrant generation and delegation algorithm

Input: Public system parameters, S_o, M_w

Output: Signcrypting warrant S_w

The original signer shall make a warrant $\boldsymbol{M_w}$ which contains the information about the type of delegation and time of delegation (t_1); it also defines the type of documents to be signcrypted by proxy signcryptor.

By using warrant $\boldsymbol{M_w}$ the original signer generates signcrypting warrant S_w by using original signer's private key $\boldsymbol{S_o}$.

1. Select $\beta \in_R Z_p$
2. $V = \beta * D_{\mathrm{h}}$
3. $S_w = (\beta - S_o . \mathcal{H}_2 (V, M_w, \mathrm{ID}_{pr}, t_1)) \bmod p$

The original signer sends (V, S_w, M_w, t_1) to proxy signcryptor.

5.5.4 Warrant verification algorithm

Input: (V, S_w, M_w)

Output: Accept or reject the signing warrant

1. The proxy signerchecks the freshness of (V, S_w, M_w, t_1).
2. On successful event, it verifies the received delegation by computing

$$V' = S_w * D_{\mathrm{h}} + \mathcal{H}_2(V, M_w, \mathrm{ID}_{pr}, t_1) P_o$$

5.5.5 Proxy signcryption algorithm

Input: Public system parameters, P_u, S_{pr}, S_w, m

Output: Signcrypted message σ

1. Select $\varphi, \mu \in_R Z_p$
2. Compute $\sigma_1 = \varphi * D_h$
3. Compute $\sigma_2 = \mu * D_h$
4. Compute $\gamma_u = \mathcal{H}_0\ (ID_u, P_u)$
5. Compute $\delta = \varphi\left(P_u + \gamma_u * M_{Pub}\right)$ (5.1)
6. Compute $\acute{K} = \mathcal{H}_3(\delta)$
7. Compute $\sigma_3 = m \oplus \acute{K}$
8. Compute $£ = \mathcal{H}_4(m)$
9. Compute $\sigma_4 = \left[S_w + \mu + (£ * S_{pr})\right] \bmod p$
10. $\sigma = (\sigma_1, \sigma_2, \sigma_3, \sigma_4, M_w)$

The proxy signcryptor uploads the signcrypted ciphertext σ on cloud.

5.5.6 Unsigncryption algorithm

Input: Public system parameters, S_u, σ

Output: Original message m *or* $\perp$

The receiver with identity ID_u will download the signcrypted ciphertext σ from cloud and perform the following operations to compute the original message m.

1. Compute $\delta' = S_u * \sigma_1$
2. Compute $\acute{K}' = \mathcal{H}_3(\delta')$
3. Compute $m' = \acute{K}' \oplus \sigma_3$
4. Compute $£' = \mathcal{H}_4(m')$
5. The receiver with Identity ID_u will then decrypt the cipher text and the message m is accepted only if following condition holds

$$\sigma_4 * D_h = V - \mathcal{H}_2(V, M_w)(P_a + \gamma_a * M_{Pub}) + \sigma_2 + £ * (P_{pr} + \gamma_{pr} * M_{Pub}) \tag{5.2}$$

otherwise it returns $\perp$

5.6 Security analysis of the proposed scheme

Proposition 1: Correctness for the generation of $\acute{K}'$. The proxy unsigncryptor first computes δ'

Let us consider the Eq. (5.1)

$$\delta = \varphi\left(P_u + \gamma_u * M_{Pub}\right)$$

Substitute $P_u = \alpha_u * D_h$ and $M_{Pub} = \vartheta * D_h$

$$\delta = \varphi\left(\alpha_u * D_h + \gamma_u * \vartheta * D_h\right)$$

$$\delta = \varphi D_h\left(\alpha_u + \gamma_u * \vartheta\right)$$

Substitute $S_u = \alpha_u + \gamma_u * \vartheta$ and $\sigma_1 = \varphi * D_h$

$$\delta = S_u * \sigma_1$$

$$\delta = \delta'$$

Then compute $\acute{K}' = \mathcal{H}_3(\delta')$

Proposition 2: Correctness for the signature. The proxy unsigncryptor checks the correctness of received signature by verifying Eq. (5.2).

$$\sigma_4 * D_h = V - \mathcal{H}_2(V, M_w)(P_a + \gamma_a * M_{Pub}) + \sigma_2 + £ * (P_{pr} + \gamma_{pr} * M_{Pub}) \tag{5.2}$$

For the correctness of Eq. (5.2), first we prove the correctness of $S_w * D_h = V - \mathcal{H}_2(V, M_w)(P_o + \gamma_a * M_{Pub})$

Let us consider LHS $S_w * D_h$

Substitute $S_w = (\beta - S_o.\mathcal{H}_2(V, M_w))$

$$= (\beta - S_o.\mathcal{H}_2(V, M_w)) * D_h$$
$$= \beta * D_h - S_o.\mathcal{H}_2(V, M_w) * D_h$$

Substitute $S_o = \alpha_o + \gamma_o * \vartheta$

$$= \beta * D_h - (\alpha_o + \gamma_o * \vartheta).\mathcal{H}_2(V, M_w) * D_h$$
$$= \beta * D_h - \mathcal{H}_2(V, M_w)(\alpha_o * D_h + \gamma_o * \vartheta * D_h).$$

Substitute $P_o = \alpha_o * D_h$ and $M_{Pub} = \vartheta * D_h$

$$= \beta * D_h - \mathcal{H}_2(V, M_w)(P_o + \gamma_o * M_{Pub}).$$

Substitute $V = \beta * D_h$

$$S_w * D_h = V - \mathcal{H}2(V, M_w)(P_o + \gamma_a * M_{Pub}) \tag{5.3}$$

= RHS

Now, Let us consider Eq. (5.2)

$$\sigma_4 * D_h = V - \mathcal{H}_2(V, M_w)(P_a + \gamma_a * M_{Pub}) + \sigma_2 + £ * (P_{pr} + \gamma_{pr} * M_{Pub})$$

Let us consider LHS $\sigma_4 * D_h$

Substitute $\sigma_4 = S_w + \mu + £ * S_{pr}$

$$= (S_w + \mu + £ * S_{pr}) * D_h$$
$$= S_w * D_h + \mu * D_h + £ * S_{pr} * D_h$$

Substitute $\mu * D_h = \sigma_2$ and $S_{pr} = \alpha_{pr} + \gamma_{pr} * \vartheta$

$$= S_w * D_h + \sigma_2 + £ * \left(\alpha_{pr} + \gamma_{pr} * \vartheta\right) * D_h$$

$$= S_w * D_h + \sigma_2 + £ * \left(\alpha_{pr} * D_h + \gamma_{pr} * \vartheta * D_h\right)$$

Substitute $P_{pr} = \alpha_{pr} * D_h$ and $M_{Pub} = \vartheta * D_h$

$$= S_w * D_h + \sigma_2 + \pounds * \left(P_{pr} + \gamma_{pr} * M_{Pub}\right)$$

Substitute Eq. 5.2 $S_w * D_h = V - \mathcal{H}_2(V, M_w)(P_o + \gamma_a * M_{Pub})$

$$= V - \mathcal{H}_2(V, M_w)(P_o + \gamma_a * M_{Pub}) + \sigma_2 + \pounds * \left(P_{pr} + \gamma_{pr} * M_{Pub}\right)$$

= RHS

Hence proved.

Theorem 1: The proposed IBPSC-MCPC-FC is secure against message disclosure attack.

Proof: A message disclosure attack is an essential aspect that has the adverse effect of disclosing sensitive EHRs during public communication. During the transmission, the proxy sends $\sigma = (\sigma_1, \sigma_2, \sigma_3, \sigma_4, M_w)$ as a signcryptext where plaintext m is hidden in σ_3 component. It must be noted that any combination of $\sigma_1 \sim \sigma_4$ by any attacker leads to unsuccessful disclosure. To successfully reveal message m, the attacker requires the ability to regenerate δ which is only possible when it is either the designated proxy server or knows the secret key S_u of the designated receiver. However, we assumed that under no circumstance, the secret key is revealed to the attacker. Therefore, the IBPSC-MCPC-FC is resistant to message disclosure attacks. ■

Theorem 2: The proposed IBPSC-MCPC-FC is secure against message modification attack.

Proof: A message modification attack appears when an adversary successfully modifies the message without noticing a designated. This impacts the transmission of the wrong EHRs to receiving end. During public transmission, the proxy sends $\sigma = (\sigma_1, \sigma_2, \sigma_3, \sigma_4, M_w)$ as a signcryptext where plaintext m is hidden in σ_3 component. It must be noted that $\sigma_1 \sim \sigma_4$ are session-dependent randomly generated components; therefore, the attacker cannot use a previously transmitted component to launch a successful attack. This is due to the hardness of the HECDP assumption. Another option for the attacker is to modify the current σ. In this case, the attacker may choose a different message m' and selects φ, μ. Therefore, it may generate $\sigma'_1, \sigma'_2, \sigma'_3$ uniformly. However, the attacker cannot generate a valid σ'_4 due to either the unavailability of S_{pr}, or the inability to extract S_{pr} from earlier transmitted signcryptext due to hardness of HECDP. Therefore, the attacker cannot launch a message modification attack. ■

Theorem 3: The proposed IBPSC-MCPC-FC is secure against proxy impersonation attack.

Proof: This aspect has a severe consequence when an attacker successfully impersonates a proxy server and transmits a message without prior knowledge of such an event at the receiver end. We may notice that a data owner delegates a warrant (V, S_w, M_w, t_1) to a designated proxy server where the owner explicitly decides who will be the designated proxy server and incorporate the respective identities during the calculation of S_w. Therefore, if an attacker tries to impersonate a valid proxy server to the receiver, it cannot reproduce a respective S_w. The reason is that such component is produced by the secret information of the data owner $S_w = (\beta - S_o.\ \mathcal{H}_2\ (V, M_w, ID_{pr}, t_1))$ mod p, and none other than the owner can delegate its rights to other proxies. Similar justification can be made to show the resistance property against user impersonation attacks due to the hardness of the HECDP assumption.

Along with the attacks mentioned above, the IBPSC-MCPC-FC protocol is safe against the delegation reply attack due to the proper use of timestamps. Therefore, the IBPSC-MCPC-FC achieves adequate security requirements.

5.6.1 Performance analysis

In this section, the security analysis of the proposed IBPSC-MCPC-FCscheme is discussed. We use the well-known AVISPA tool [31,32] to discuss the security proof and demonstrate that the proposed scheme is not susceptible to replay and man-in-the-middle attack. It should be noted that for any security protocol, AVISPA only handles replay and man-in-the-middle threats against an attacker.

The HLPSL [33] code is written for the proposed scheme with the different roles like original signer, proxy signer and trusted third party. This code is then executed using SPAN and AVISPA with the backends OFMC and CL-AtSe. We can see that no attacks were discovered by OFMC. In other words, for a limited number of sessions as specified in the role of the environment, the stated security goals were achieved. The proposed protocol is also executed with CL-AtSe backend for bounded number of sessions. The output shows that the protocol is safe under CL-AtSe also. The software resources such as Oracle VM Virtual Box and Security protocol animator (SPAN) are used. The output of AVISPA is shown in Figs. 5.2 and 5.3.

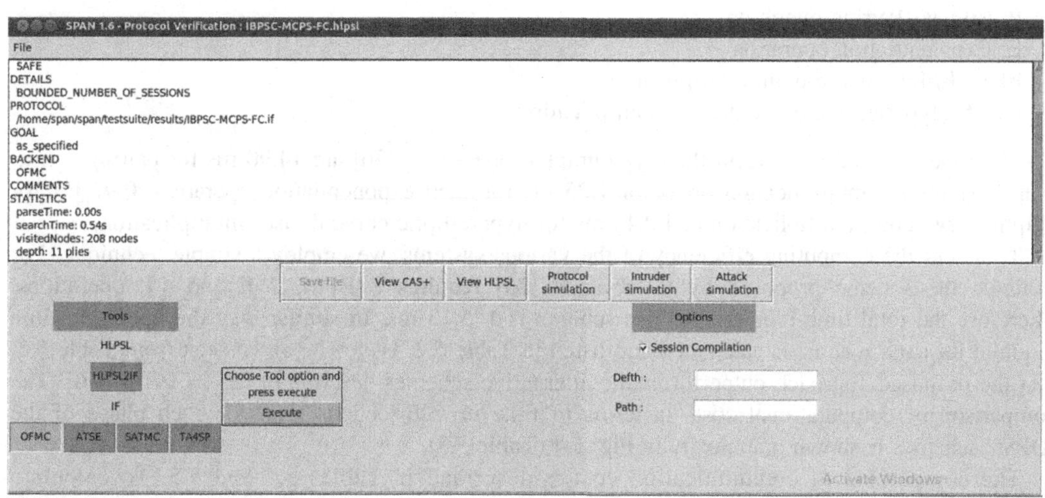

FIGURE 5.2

OFMC output.

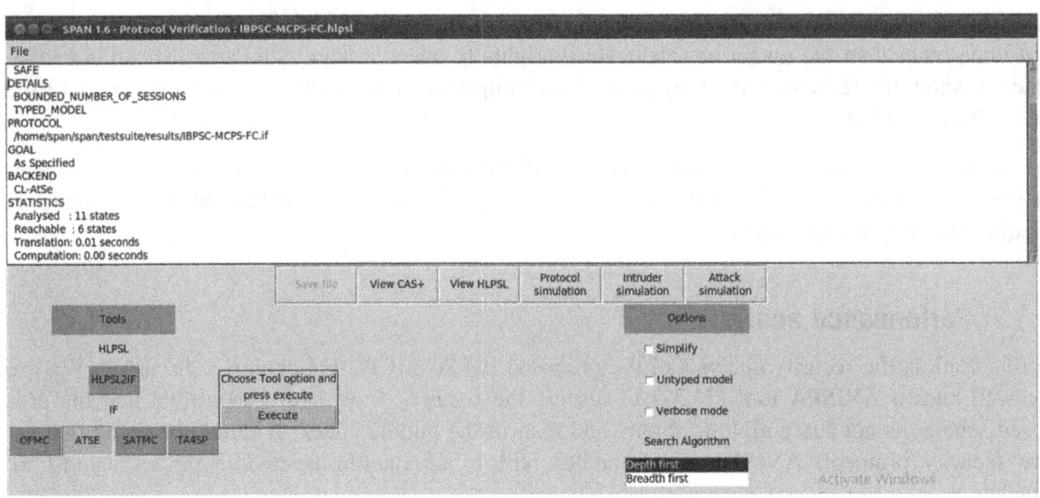

FIGURE 5.3

CL-AtSe output.

We have done the comparison of our proposed IBPSC-MCPC-FCscheme with the easting proxy signcryption schemes, which include G. Swapna et al. [25], Y. Ming et al. [34], C. Zhou et al. [27], H. Yu et al. [35], Guo and Deng [29], Hundera et al. [36]. The comparison outcomes are listed in Table 5.2. We define some notations as follows:

BPM: Bilinear Pairing multiplications
P: Bilinear Pairing operation
E: Exponentiation operation
EPM: Elliptic curve point multiplication
HDM: Hyperelliptic curve divisor multiplication.

The time required to perform the cryptographic operations [36] are 14.90 ms for pairing operation, 4.31 ms for multiplication operation,1.25 ms for each exponentiation operation, 0.97 ms for elliptic curve point multiplication and 0.48 ms for hyperelliptic curve divisor multiplication.

To assess the computing efficiency of the various systems, we employ a simple technique. For example the scheme proposed by G. Swapna [25] requires 8 BPM, 1 E and 4 P operations. Therefore the total time required for this scheme is 125.13 ms. In similar way the operation time required for each scheme is calculated and listed in Table 5.2. Hence it can be seen from Table 5.2, that the proposed approach outperforms the alternative schemes describe in [25,27,29,34–36]. The comparison of computational costs in terms of time in milliseconds (ms) for each phase of the IDPSC schemes is shown graphically in Fig. 5.4 (Table 5.3).

The comparison of communication cost is described in Tables 5.4 and 5.5. To calculate the communication cost we have considered that a single hash value ($\mathcal{H}$) is communicated it takes 512 bits, a message ($\mathcal{M}$) is considered to be of 2048 bits and a pairing operation ($\mathfrak{E}$) is considered to be of 1024 bits, Q is considered to be of 160 bits and N is considered to be of

Table 5.2 Computational cost comparison.

Scheme	Warrant generation and delegation	Proxy secret key generation/ warrant verification	Proxy signcryption	Unsigncryption	Total
G. Swapna [25]	3 BPM	2 P	3 BMP + 1 E + 1 P	2 BPM + 3 P	8 BPM +1 E + 4 P
Y. Ming [34]	3 BPM	2 P	3 BPM +4 P + 3 E	4 M + 6 P	6 BPM + 7 E + 12 P
C. Zhou [27]	3 BPM	2 P + 1 BPM	3 BPM + 1 P + 1 E	4 BPM + 4 P	11 BPM + 1 E + 7 P
H. Yu [35]	1 M	1 M	1 BPM + 3 E + 1 P	1 BPM + 1 E + 5 P	2 BPM + 6 E + 6 P
H. Guo [29]	1 EPM	3 EPM	4 EPM	6 EPM	14 EPM
N. Hundera [36]	3 BPM	1 BPM + 2 P	4 BPM + 2 P	2 BPM + 4 P	10 BPM + 8 P
Ours	2 HDM	2 HDM	5 HDM	6 HDM	15 HDM

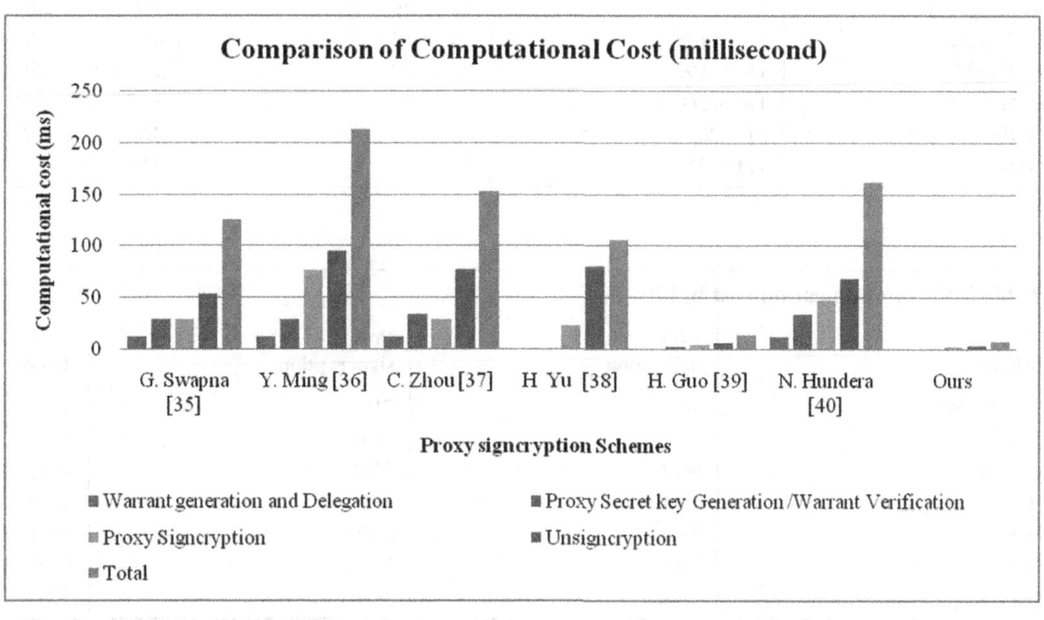

FIGURE 5.4

Comparison of computation cost of alternative schemes with proposed IBPSC-MCPC-FC scheme.

Table 5.3 Computational cost (millisecond).

Scheme	Warrant generation and delegation	Proxy secret key generation/ warrant verification	Proxy signcryption	Unsigncryption	Total
G. Swapna [25]	12.93	29.8	29.08	53.32	125.13
Y. Ming [34]	12.93	29.8	76.28	94.4	213.41
C. Zhou [27]	12.93	34.11	29.08	76.84	152.96
H. Yu [35]	1.25	1.25	22.96	80.06	105.52
H. Guo [29]	0.97	2.91	3.88	5.82	13.58
N. Hundera [36]	12.93	34.11	47.04	68.22	162.3
Ours	0.96	0.96	2.4	2.88	7.2

Table 5.4 Communication cost comparison.

Scheme	Proxy delegation	Proxy signcryption	Total
G. Swapna [25]	$1\mathcal{M} + 2\mathbb{G}$	$2\mathcal{M} + 3\mathbb{G}$	$3\mathcal{M} + 5\mathbb{G}$
Y. Ming [34]	$1\mathcal{M}$	$1\mathbb{G} + 2\mathcal{H}$	$1\mathcal{M} + 1\mathbb{G} + 2\mathcal{H}$
C. Zhou [27]	$3\mathcal{M} + 1\mathcal{H}$	$1\mathcal{M} + 1\mathbb{G} + 2\mathcal{H}$	$4\mathcal{M} + 1\mathbb{G} + 3\mathcal{H}$
H. Yu [35]	$1\mathcal{M} + 1\mathcal{H}$	$1\mathbb{G} + 2\mathcal{H}$	$1\mathcal{M} + 1\mathbb{G} + 3\mathcal{H}$
H. Guo [29]	$1\mathcal{M} + 2\mathcal{Q}$	$2\mathcal{M} + 5\mathcal{Q}$	$3\mathcal{M} + 7\mathcal{Q}$
N. Hundera [36]	$\mathcal{M} + 2\mathbb{G}$	$2\mathcal{M} + 1\mathbb{G} + 1\mathcal{H}$	$3\mathcal{M} + 3\mathbb{G} + 1\mathcal{H}$
Ours	$1\mathcal{M} + 2\mathrm{N}$	$2\mathcal{M} + 5\mathrm{N}$	$3\mathcal{M} + 7\mathrm{N}$

Table 5.5 Communication cost in bits.

Scheme	Proxy delegation	Proxy signcryption	Total
G. Swapna [25]	4096	7168	11264
Y. Ming [34]	2048	1184	3232
C. Zhou [27]	6656	3232	9888
H. Yu [35]	2560	1184	3744
H. Guo [29]	2368	6656	9024
N. Hundera [36]	4096	5200	9296
Ours	2208	4896	7104

80 bits. Fig. 5.5 shows the details of communication cost comparison of various schemes with proposed scheme in bits. Hence it can be seen that the proposed approach outperforms the alternative schemes.

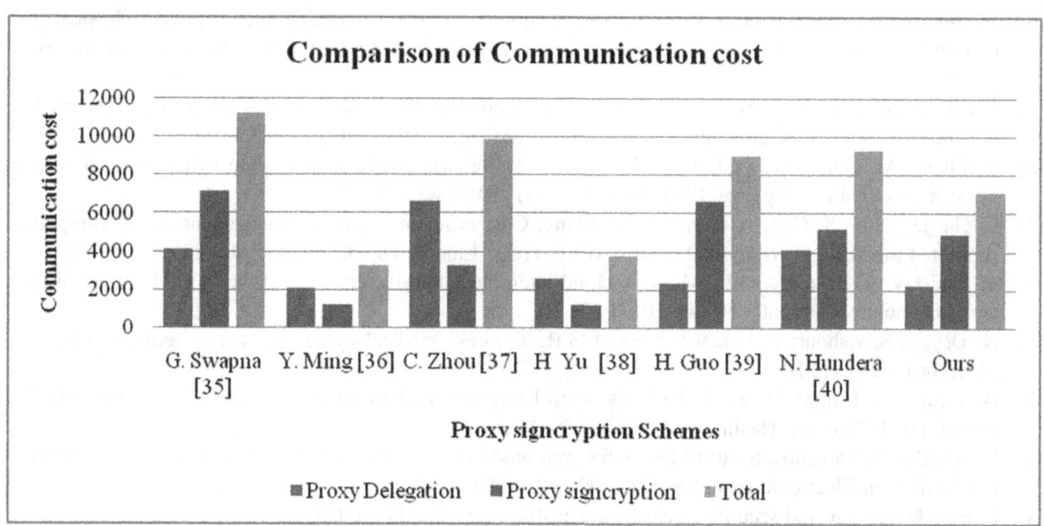

FIGURE 5.5

Comparison of communication cost of alternative schemes with proposed IBPSC-MCPC-FCScheme.

5.7 Conclusion

Despite the supremacy of MCPS, the insecure and high-latency links between cloud data centers and medical equipment are a source of concern in the practical usage of the MCPS. Fog computing and MCPS integration promise to be a possible way to solve this issue. This chapter explains how to design a safe Identity-based Proxy Signcryption and apply it to Fog-based MCPS. Under the hardness of the HECDP assumption, the proposed method meets sufficient security criteria. The revised protocol is safe in practice, according to the simulation research conducted with AVISPA tools. Furthermore, a thorough analysis of its performance reveals its efficiency in terms of processing and transmission expenses.

References

[1] H. Landi, Healthcare, tech companies are vying for a piece of back-to-work market. Here's are the new opportunities post-COVID, 2020. Fiercehealthcare.com, pp. na-na.

[2] I. Keshta, A. Odeh, Security and privacy of electronic health records: concerns and challenges, Egypt. Inform. J. 22 (2) (2021) 177–183.

[3] M.R. Fuentes, Cybercrime and other threats faced by the healthcare industry. Trend Micro, 2017.

[4] M.V. Fontanilla, Cybercrime pandemic, Eubios J. Asian Int. Bioeth. 30 (4) (2020) 161–165.

[5] M.M. Nair, A.K. Tyagi, R. Goyal, Medical cyber physical systems and its issues, Proc. Comp. Sci. 165 (2019) 647–655.

[6] J.I. Jimenez, H. Jahankhani, S. Kendzierskyj, Health care in the cyberspace: medical cyber-physical system and digital twin challenges, Digital Twin Technologies and Smart Cities, Springer, Cham, 2020, pp. 79–92.

[7] J.H. Kim, A review of cyber-physical system research relevant to the emerging IT trends: industry 4.0, IoT, big data, and cloud computing, J. Ind. Integr. Manag. 2 (03) (2017) 1750011.

[8] Y. Zhang, M. Qiu, C.W. Tsai, M.M. Hassan, A. Alamri, Health-CPS: healthcare cyber-physical system assisted by cloud and big data, IEEE Syst. J. 11 (1) (2015) 88–95.

[9] L. Gu, D. Zeng, S. Guo, A. Barnawi, Y. Xiang, Cost efficient resource management in fog computing supported medical cyber-physical system, IEEE Trans. Emerg. Top. Comput. 5 (1) (2015) 108–119.

[10] T. Sanislav, S. Zeadally, G.D. Mois, A cloud-integrated, multilayered, agent-based cyber-physical system architecture, Computer 50 (4) (2017) 27–37.

[11] N. Dey, A.S. Ashour, F. Shi, S.J. Fong, J.M.R. Tavares, Medical cyber-physical systems: a survey, J. Med. Syst. 42 (4) (2018) 1–13.

[12] B. Tiwari, A. Kumar, Role-based access control through on-demand classification of electronic health record, Int. J. Electron. Healthc. 8 (1) (2015) 9–24.

[13] C. Ilioudis, G. Pangalos, Security issues for web based electronic health care record, in: Third European Conference on Electronic Health Records (EuroRec'99), 1999.

[14] S. Brands, Privacy and security in electronic health, Pricewaterhouse Coopers CCE J. (2003).

[15] P.V. Devi, V. Kalaichelvi, Security issues in medical cyber physical systems (MCPS)—a survey, Int. J. Pure Math. 117 (20) (2017) 319–324.

[16] L. Lee, Y. Liu, Distributed access control based on proxy signature in medical cyber-physical systems, Recent. Pat. Computer Sci. 10 (3) (2017) 194–200.

[17] P. He, K. Xue, J. Yang, Q. Xia, J. Liu, D.S. Wei, FASE: fine-grained accountable and space-efficient access control for multimedia content with in-network caching, IEEE Trans. Netw. Serv. Manag. (2021).

[18] S. Deore, R. Bachche, A. Bichave, R. Patil, Review on applications of blockchain for electronic health records systems, International Conference on Image Processing and Capsule Networks, Springer, Cham, 2021, pp. 609–616.

[19] R.Y. Patil, S.R. Devane, Hash tree-based device fingerprinting technique for network forensic investigation, Advances in Electrical and Computer Technologies, Springer, Singapore, 2020, pp. 201–209.

[20] G. Pattewar, N. Mahamuni, H. Nikam, O. Loka, R. Patil, Management of IoT devices security using blockchain—a review, Sentimental Anal. Deep. Learn. (2022) 735–743.

[21] X. Li, K. Chen, Identity based proxy-signcryption scheme from pairings, in: IEEE International Conference onServices Computing, 2004. (SCC 2004). Proceedings, 2004, September, pp. 494–497. IEEE.

[22] Q. Wang, Z. Cao, Efficient ID-based proxy signature and proxy signcryption form bilinear pairings, International Conference on Computational and Information Science, Springer, Berlin, Heidelberg, 2005, pp. 167–172.

[23] M. Wang, H. Li, Z. Liu, Efficient identity based proxy-signcryption schemes with forward security and public verifiability, International Conference on Networking and Mobile Computing, Springer, Berlin, Heidelberg, 2005, pp. 982–991.

[24] H.M. Elkamchouchi, Y. Abouelseoud, A new proxy identity-based signcryption scheme for partial delegation of signing rights, IACR Cryptol. ePrint Arch. 2008 (2008) 41.

[25] G. Swapna, P.V.S.S.N. Gopal, T. Gowri, P.V. Reddy, An efficient ID-based proxy signcryption scheme, Int. J. Inf. Netw. Sec. 1 (3) (2012) 200.

[26] H.Y. Lin, T.S. Wu, S.K. Huang, Y.S. Yeh, Efficient proxy signcryption scheme with provable CCA and CMA security, Comp. Math. Appl. 60 (7) (2010) 1850–1858.

[27] C. Zhou, Y. Zhang, L. Wang, A provable secure identity-based generalized proxy signcryption scheme, Int. J. Netw. Secur. 20 (6) (2018) 1183–1193. Available from: https://doi.org/10.6633/IJNS.201811206.18.

[28] Y. Huifang, W. Zhicang, L. Jianmin, G. Xinzhe, Identity-based proxy signcryption protocol with universal composability, Sec. Commun. Netw. 2018 (2018). Available from: https://doi.org/10.1155/2018/9531784. Article ID 9531784.

[29] H. Guo, L. Deng, An identity based proxy signcryption scheme without pairings, Int. J. Netw. Secur. 2 (4) (2020) 561–568.

[30] J. Wu, Identity-based proxy signcryption schemes, Applied Mechanics and Materials, Vol. 380, Trans Tech Publications Ltd., 2013, pp. 2605–2608.

[31] P.R. Yogesh, Formal verification of secure evidence collection protocol using BAN logic and AVISPA, Proc. Comp. Sci. 167 (2020) 1334–1344.

[32] R.Y. Patil, S.R. Devane, Network forensic investigation protocol to identify true origin of cyber crime, J. King Saud. Univ. Comp. Inf. Sci. (2019).

[33] D. Von Oheimb, The high-level protocol specification language HLPSL developed in the EU project AVISPA, in: Proceedings of APPSEM 2005 Workshop, 2005, September, pp. 1–17.

[34] Y. Ming, Y. Wang, Proxy signcryption scheme in the standard model, Sec. Commun. Netw. 8 (8) (2015) 1431–1446.

[35] H. Yu, Z. Wang, Construction of certificateless proxy signcryption scheme from CMGs, IEEE Access. 7 (2019) 141910–141919.

[36] N.W. Hundera, Q. Mei, H. Xiong, D.M. Geressu, A secure and efficient identity-based proxy signcryption in cloud data sharing, KSII Trans. Internet Inf. Syst. (TIIS) 14 (1) (2020) 455–472.

CHAPTER

Secure medical image storage and retrieval for Internet of medical imaging things using blockchain-enabled edge computing

6

Vijay Jeyakumar[1], K. Rama Abirami[2], S. Saraswathi[3], R. Senthil Kumaran[4] and Gurucharan Marthi[5]

[1]Department of Biomedical Engineering, Sri Sivasubramaniya Nadar College of Engineering, Chennai, Tamil Nadu, India [2]Department of Information Science and Engineering, Dayananda Sagar Academy of Technology and Management, Bengaluru, Karnataka, India [3]Department of Computer Science and Engineering, Sri Sivasubramaniya Nadar College of Engineering, Chennai, Tamil Nadu, India [4]Zoho Corporation Private Limited, Chennai, Tamil Nadu, India [5]Department of Neurology and Neurosurgery, McGill University, Montreal, QC, Canada

6.1 Introduction

The advancement of the healthcare sector is a matter of great concern to most countries of the world as this is directly related to the lives and well-being of their citizens. Most countries have been increasing their annual funding for research and development in the healthcare sector. Healthcare research must be a continuous exercise as this would help in ameliorating the quality of living and prevent the spread of diseases. With the recent advancements in the field of science and technology, there is a scope for improvement in healthcare by integrating the existing medical systems with the latest computer technologies. This would highly benefit the medical specialists and surgeons in diagnosing and curing several diseases.

Most of the medical problems that occur within the human body are not visible to the naked eye and hence would need techniques that pass through the layers of skin to visualize the internal organs of the body for diagnosis and treatment. Medical imaging techniques enable us to see the inside of the human body and help us to monitor, diagnose, and treat several medical problems. Ever since its introduction, medical imaging has become an important part of the healthcare sector and healthcare research, allowing medical practitioners and researchers to understand more about the human body and its internal organization. The field has opened a new avenue in the diagnosis and treatment of diseases, and several research works are being carried out to develop new and innovative methods for the advancement of the techniques.

Early detection of diseases is vital in saving lives, which can be achieved with the integration of advanced techniques into the existing healthcare domain. Medical imaging is extremely helpful in diagnosing cancer as it allows doctors to detect tumors at a very early stage. It has been observed

Intelligent Edge Computing for Cyber Physical Applications. DOI: https://doi.org/10.1016/B978-0-323-99412-5.00004-6

that with the advancement in medical imaging methods, such as mammography, there has been a 30% decrease in deaths involving breast cancer. Most of the common medical imaging modalities used today are summarized in Fig. 6.1.

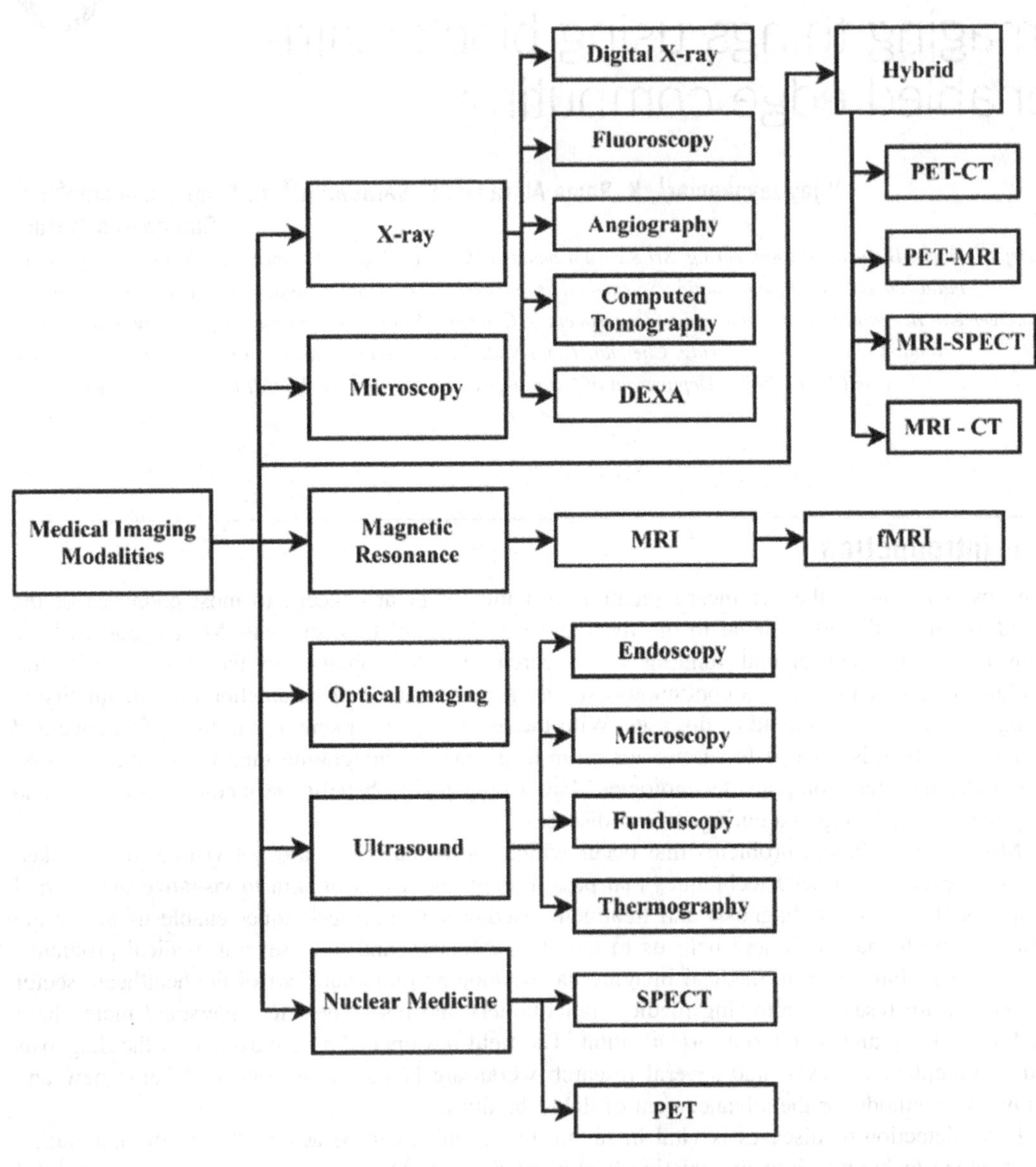

FIGURE 6.1

Medical imaging modalities.

6.1.1 Types of imaging modalities

6.1.1.1 X-Ray

X-Ray is a type of electromagnetic radiation that uses electromagnetic waves to form images of the inside of our body. It is one of the oldest techniques of medical imaging, which is widely used even today. Detection of bone fractures is the most common application of X-Ray. Dental X-Rays are a major source of medical imaging information for dentists, aiding them in the diagnosis and treatment of dental problems.

6.1.1.2 Computed tomography

Computed tomography (CT) is an advanced version of X-Ray that uses multiple X-Rays to form a 3D visualization of the internal parts to be diagnosed. CT provides information on the hard and soft tissues along with the blood vessels that provide details on the location of the infection whereas X-Rays provide only 2D information on the hard tissues, which helps in identifying abnormalities, if there are any. CT of the brain is most commonly used to diagnose severe brain injuries or to understand the nature of the brain stroke.

6.1.1.3 Magnetic resonance imaging

Magnetic resonance imaging (MRI) is a medical imaging technique that makes use of radio waves and magnetic fields to produce high-quality 3D images of the internal tissues and organs of the human body. The strong magnetic field is introduced and is used to construct high-quality images of the tissues in the body. Doctors and patients benefit strongly from the brain and spinal cord imaging from MRI scans, and the MRI technique is extensively used as it does not involve any harmful ionizing radiation as in CTs and X-Rays.

6.1.1.4 Ultrasound modalities

Ultrasound makes use of sound waves to obtain medical images—a probe is kept on the skin of the patient after which high-frequency sound waves are transmitted and reflected in the form of an image. This technique is widely used in visualizing the images of the fetus in the body and monitoring the baby's growth during pregnancy.

6.1.1.5 Positron emission tomography

Positron emission tomography (PET) uses a special dye of radioactive tracers, which is injected into the vein closer to the body part to be diagnosed. The internal organs and tissues may absorb this radiotracer at different amounts, which is detected by the PET scanner.

6.1.1.6 Hybrid modalities

In addition to these individual medical imaging procedures, hybrid modalities have been developed, which combine two or more imaging modalities such as PET + CT or ultrasound + CT. Such fusion imaging have several advantages, such as a more precise evaluation of the disease and improved diagnostic confidence as the clinicians can validate a particular lesion or problem with multiple modalities.

6.1.2 Picture archiving and communication system

Due to the rapid growth in communication technology, multimedia data in various applications are increasing enormously in different sectors, such as entertainment, education, business, healthcare, and security. An efficient system for storing and retrieving the data in a systematic manner is an important requirement. During the past five decades a lot of research have been carried out in the dimension of digital data storage and retrieval [1]. Picture archiving and communication system (PACS) is a technique in the medical imaging sector that is used by several healthcare institutions to store, retrieve, and display medical images in electronic format. This system consists of several computer workstations in a healthcare institution, which can be accessed by various medical practitioners such as radiologists and physicians. PACS helps in transmitting the medical images from the image acquisition site to a server that can be accessed from anywhere. The implementation of PACS in hospitals has eliminated the use of film jackets and conventional films to store medical images. Additionally, accessing medical imaging data with multiple modalities simultaneously at several locations has become convenient with the deployment of PACS in hospitals.

6.1.2.1 Digital imaging and communications in medicine

PACS make use of the DICOM (digital imaging and communications in medicine) standard to store and retrieve medical images from their database. DICOM is considered a universal medical imaging format with most medical popular imaging modalities having DICOM protocol as their file format. This ensures that the data from all scanners would be in an organized manner and can be exchanged across several PACS workstations and healthcare institutions, promoting better collaboration between radiologists. Recently, cloud-based PACS have been gaining importance with several hospitals migrating to cloud servers to store their large medical databases.

6.1.2.2 PACS workflow

PACS consists of four major modules as given in Fig. 6.2: (1) Image acquisition systems, (2) Communication network, (3) PACS database and server, and (4) Integrated workstations for display [2]. Medical images are scanned from the acquisition systems, such as CT, X-Ray, and MRI, which are then transmitted to the PACS archive for image storage and retrieval. The transmission of images takes place with the help of secure and wired communication networks (e.g. telephone networks, fiber optical cables, etc.) or wireless communication networks (e.g. cell phone networks, satellite communication networks, microwave networks, etc.), which are then stored on the PACS server. Medical professionals are provided with workstations or computer systems that are connected to the PACS servers through wireless networks. Medical images are retrieved from the PACS server on request from clinicians who use them to diagnose certain conditions from the images. Several PACS software also provide basic image processing commands, such as zoom in, crop, and image enhancement, which enable doctors to understand the medical condition better.

6.1.2.3 PACS uses

The implementation of PACS in hospitals has enabled efficient imaging data management and easier access to medical reports and medical images. With the computerized software bundled with the PACS database, radiologists can zoom in and get a detailed view of the location, which aids in a better diagnosis of the disease concerned. The imaging data can be accessed from the medical

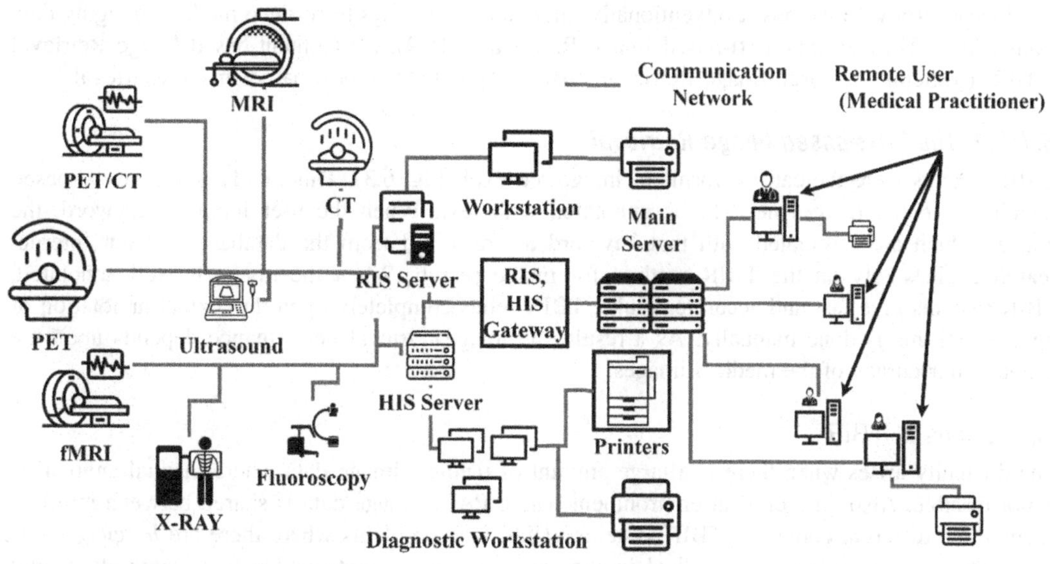

FIGURE 6.2

PACS workflow.

specialist's computer through a centralized PACS system, integrating various departments of the hospital.

Since PACS uses modern cloud-based systems, it is economically beneficial to healthcare institutions as it cuts down the cost of large digital storage devices. In combination with other forms of patient data, such as the electronic health record (EHR), radiology information system (RIS), and clinical information system (CIS), PACS organizes existing patient information with the medical images in chronological order, which can be easily retrieved for diagnosis by the clinician or doctor [3]. The user-friendly software of PACS coupled with effective management of patient data provides significant benefits to the healthcare institutions.

6.1.3 Retrieval of medical images in PACS

The medical images that are stored in the PACS need to be retrieved at several stages by different clinicians in the hospital setting for various purposes. This fetching of the medical images from the PACS archive through wired/wireless communication networks is called Medical Image Retrieval. This form of retrieving medical images is challenging in nature due to the multimodal, multidirectional and multianatomical nature of medical images. With the increase in the amount of medical data, several efficient and novel image retrieval methods have been developed that help manage such huge imaging data. The recent boom in computerized technologies has aided in developing a few image retrieval methods. Conventionally, there are three ways to retrieve medical imaging data from PACS. They are (1) Text-based Image Retrieval (TBIR), (2) Content-based Image Retrieval (CBIR) (3) Semantic-based Image Retrieval (SBIR) [4] and (4) Hybrid-based Image Retrieval.

6.1.3.1 The Text-based Image Retrieval

TBIR systems were the earliest forms of image retrieval (Fig. 6.3). This involves a keyword-based search where the images need to be annotated with text. When the user inputs a keyword, the images which are associated with that keyword are retrieved from the database. Popular internet search engines rely on the TBIR method for image search. When the image is well annotated, TBIR systems give fast and accurate results. TBIR relies completely upon the textual annotation of images and this is done manually. As a result, the image retrieval performance depends upon the annotation accuracy of the medical images.

Limitations of TBIR

The difficulty arises when there is a large amount of medical image data where manual annotation is not feasible. Also, in a clinical environment where PACS image data is shared between multiple hospitals in different countries, TBIR systems would face problems where there are *language barriers*. The difference between medical images is often very minute and a few keywords cannot describe its rich information in detail [5]. Hence, CBIR systems were introduced that addressed these issues faced by the existing TBIR systems.

6.1.3.2 Content-based Image Retrieval systems

CBIR is a rapidly developing field of interest that is beneficial to the PACS environment in hospitals. In this system, the medical images are retrieved from the large database based on their visual content instead of manually annotated keywords. The main goal of the CBIR systems is to perform effective feature extraction from medical images. The visual features such as shape, color, size and edges form the main criteria as features to be extracted are known as low-level features. These features are picked up by the descriptors such as convolution layers and are stored in a database as feature vectors. When a query image is given as an input to the PACS, similar features are

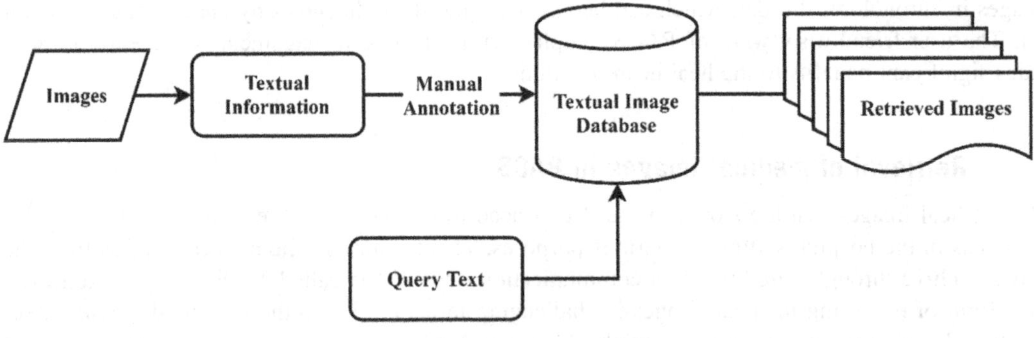

FIGURE 6.3

Text-based Image Retrieval.

extracted from the query input and are processed. Based on similarity metrics such as Euclidean or Manhattan distances, color histograms and mathematical transforms, the query image features are compared to the features of those in the database. Based on the similarity scores, the medical images are retrieved from the database and displayed to the user. Content-based Medical Image Retrieval Systems (CBMIR) is a growing subset of the CBIR systems, where several different models are being developed at various research organizations.

Limitations of CBIR

The extraction of these low-level features from medical images brings about a semantic gap as the high-level features from the images are left out and cannot be extracted by these feature descriptors.

These are overcome by the SBIR methods. CBIR Architecture is given in Fig. 6.4.

6.1.3.3 Semantic-based Image Retrieval

SBIR systems aim to bridge the semantic gap that occurs between the low-level and high-level features in the medical images. The high-level features are defined as the human perception of images. In medical images, several internal anatomical parts of the body are quite similar and this has a risk of developing the “semantic gap". SBIR systems incorporate a combination of the CBIR systems that extract low-level features from medical images along with high-level semantics to retrieve medical images from databases. Once the low-level features of the images are extracted, the semantic description of the medical images are extracted from the images and are stored as high-level feature keywords. A semantic mapping takes place that will choose the best combination for effective medical image retrieval. There are several types of SBIR systems that intend to reduce the “semantic gap” by combining different techniques (Fig. 6.5).

6.1.3.4 Hybrid-based image retrieval

Apart from these major image retrieval techniques, several other methods are being researched. The advancement of Deep Learning techniques involving Convolutional Neural Networks (CNN) has rapidly advanced the research on CBIR systems [6]. Relevance feedback image retrieval is a technique that allows the user to evaluate the medical image retrieval results as a result of which

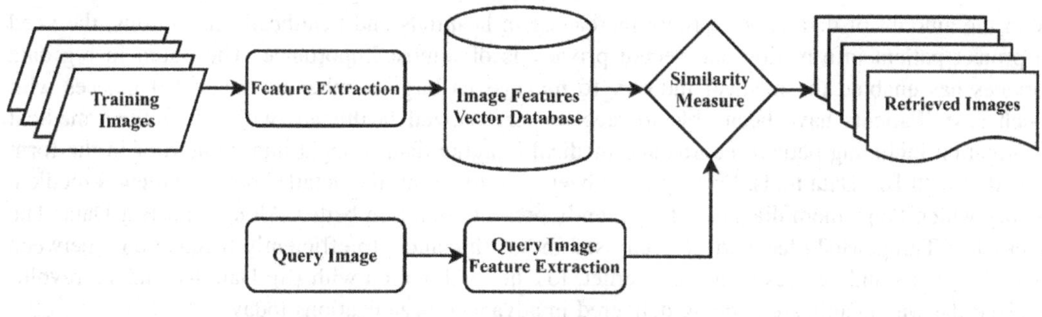

FIGURE 6.4

Content-based Image Retrieval.

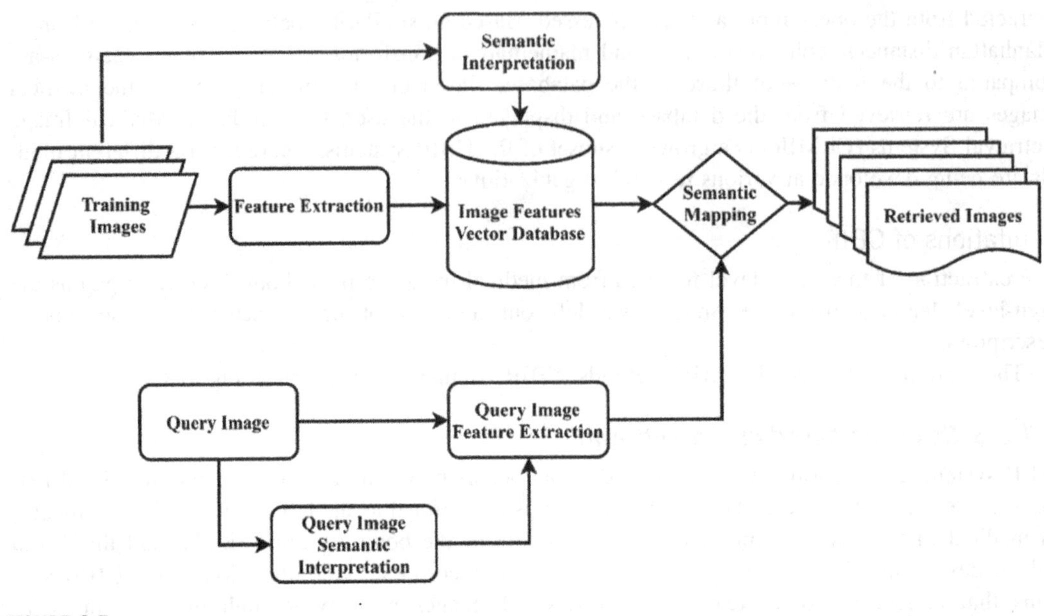

FIGURE 6.5

Semantic-based Image Retrieval.

the subsequent retrieval results are refined. Annotation based image retrieval (ABIR) is another retrieval method that reduces the semantic gap by using automatic image annotation along with image segmentation. Active research is being undertaken in developing such Medical Image Retrieval methods [7] Deep Learning frameworks, CNNs, Kernel-based retrieval, and Sketch-based retrieval are some of the recent methods that have been developed in medical image retrieval systems.

6.1.4 Secure storage mechanisms

With the amount of data exponentially increasing in hospitals and healthcare institutions, the need to protect patient information and secure privacy is of utmost importance. The boom in big data services has enabled the medical industry to manage its large number of medical databases with much ease. Patients have been able to receive personalized healthcare with all of their medical information including patient records and medical imaging data brought under one roof in the form of PACS with Big Data tools. Doctors have been able to get all the details about a patient's medical history which helps them diagnose the patient's present condition better. Along with Big Data, The Internet of Things (IoT) has widely been used in health centers to efficiently manage data between several systems and devices over the Internet. IoT in combination with Big Data has indeed revolutionized the way healthcare is being delivered in advanced organizations today.

Blockchain Technology is a device management system that permits the secure transmission of data without the involvement of any third-party system [8]. In combination with IoT, Blockchain

Technology enhances the stability and privacy of such medical data. Using this, device authentication can be performed which can prevent the risk of cyber-attacks or any malicious activity. Cloud Computing is yet another technology that is being incorporated into healthcare systems. It is used to replace huge data servers in hospitals by storing all patient data and medical imaging data on the cloud which can later be accessed by PACS and image retrieval techniques.

By combining all the above-mentioned technologies, a secured storage mechanism can be developed with Blockchain Technology and Cloud Computing services to effectively manage and retrieve medical imaging data from large PACS databases in hospitals and healthcare institutions.

6.2 Internet of Things

In the rapidly evolving world of the Internet which bridges the gap between a particular individual and vastly found information, the Internet of Things (IoT) which happens to intertwine with the confounded internet consists of a collaborated bonding between magnitudes of digital and mechanical action, embedded resources and a network where data can be transmitted, and an enormous quantity of physical devices where data is accessible through the network.

6.2.1 Internet of medical things

In the past years, IoT has enormous advancements in the field of healthcare. IoT is helping patients to lead safe and happy life by providing them with formidable care and support. Moreover, it has made the Doctor-Patient interaction more unchallenging and affordable. IoT devices keep us from visiting doctors often because healthy life is possible just by monitoring the patient's health by remote monitoring which provides a keen record of the patient's well-being. This also helps us prevent more complications in future so that we can avoid unwanted admissions to the hospital. IoT devices play an essential role in hospitals by furnishing medical equipment and machinery which also include devices that come along with sensors to locate defibrillators, oxygen pumps and other hospital equipment.

Maintaining optimum health is highly essential to stay happy and cheerful. Certainly, both mental and physical health are counted together for contented well-being. The role of the Internet of Medical Things (IoMT) has reliably made it possible for every being to achieve this goal of staying healthy. By interconnecting healthcare and digital resources, limitless data is being transmitted for the working of a specific application. The augmentation and growth of healthcare seem to be phenomenally upgrading constantly [9]. The application of IoMT devices includes implantable glucose monitoring machines, activity trackers for cancer treatments, medical alert systems, wireless sensors, medication dispensers, etc. The IoMT Ecosystem is shown below in Fig. 6.6.

6.3 Edge computing

Computer power is used in almost all the sectors, education, agriculture, healthcare, transport, payment systems, R&D in all sectors etc. Data is the most vital thing and storage of that

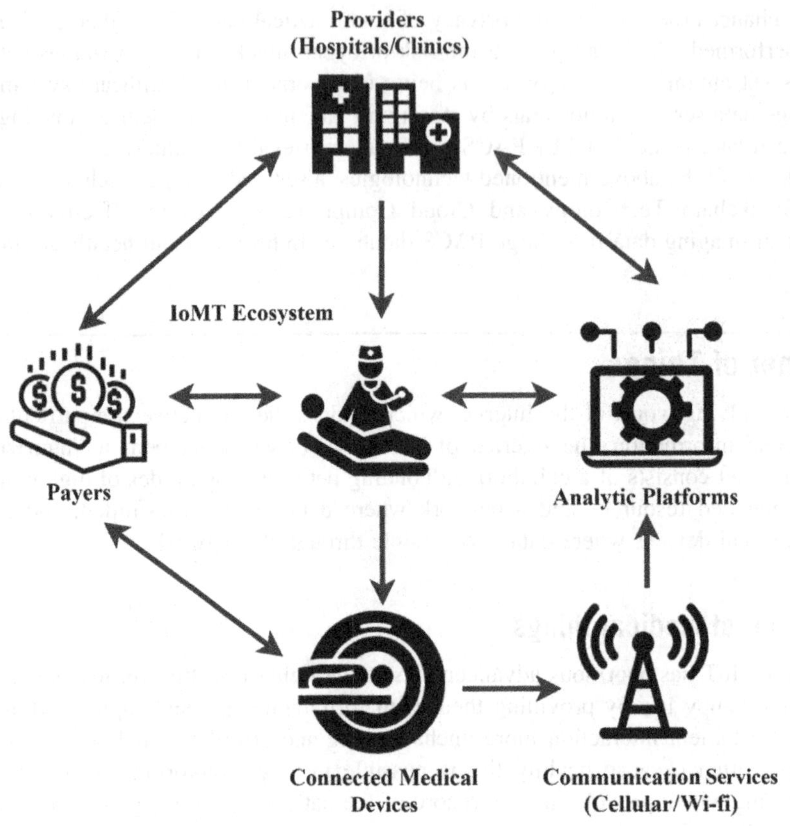

FIGURE 6.6

The IoMT Ecosystem of Dynamic collaborations.

data & retrieval of the computed data in a secure manner is the ultimate goal which backed the evolution of computing from mainframe to cluster to the grid to cloud computing. Now the buzz-word is moving from cloud computing to edge computing to cover the blind spot of cloud computing. This edge computing idea is already achieved by most of the Network Monitoring & Operation management tools (ITOM) by running their small agent at the customer network which sends computed data and alarms to their cloud server.

Edge computing is capturing, processing, and analyzing the data near the data sources to improve the latency even in slow network areas. Computed data from the edge server then be pushed into the cloud server periodically. This computing at the data place is much required for those sectors producing numerous data in a second. Here edge computing solves the key challenges of bandwidth, latency, resilience over network fluctuations, and data authority. Table 6.1 compares the various computing technology frameworks that are used today for handling big data.

Table 6.1 Comparison of various computing technologies.

Technology	Functions	Advantages	Challenges
Mainframe/ local server Computing	Powerful computing machine within the Organization network	It can store huge data & fast retrieval.	High cost
Cluster Computing	The cluster of servers replaces the costly mainframe server & shares the processes	Horizontally scalable by adding new nodes	Restricted to geography location.
Grid Computing	Geo distribute internet-connected systems to share the computing job	Cheap, efficient, fail-safe, load sharing etc.	Non-interactive job, difficult to process Memory-intensive works, decentralized management Troubles licensing
Cloud Computing	Almost similar to Grid but differs in architecture. Remote servers are hosted on the internet to do all computing work.	Reduced cost, increased return. Single domain, safe data storage. Easy access from anywhere through the internet	Internet-dependent, denial of service possibility, security & technical complexities.
Edge Computing	Edge server at/near the data source which connected with cloud server	Computes at data sources and reports can be generated in low network areas. Suitable for real-time processes.	Data Accumulation, distributed computing

6.3.1 Salient features of edge computing

- It opens a new way to implement Artificial Intelligent, Machine Learning and other technologies into the edge server, and edge networks to process the huge medical image data quickly & efficiently. This can result in the reduction of redundant data in cloud storage. Moreover, implementation of an automated computer-aided diagnosis in the server is possible and more efficient.
- As the server/data center is very close to the medical data sources, implementing high security for the data is much easier than achieving cloud data security. Easily design the connection only to the specific devices and can restrict the others.
- It helps to reduce the network bandwidth as the redundant data is trimmed at the edge server and the computed data is sent to the cloud server.
- Easy to maintain the network connection between the data sources & the edge servers which nullifies the data loss. Also, the edge servers can connect and send the computed data to the cloud whenever the network is available well in a periodic manner.
- Highly scalable & reliable.

6.3.2 Role of edge computing in healthcare

The healthcare system covers various works like patient treatment, doctors' availability management, ambulance, hospital maintenance, hospital device management, insurance etc. To manage them well, many hospitals are using specialized software based on their needs. Nowadays those hospitals want to manage them from anywhere and so switch from on-premises software systems to the cloud.

Medical device vendors are also in line with current trends and support their device software to produce the data and send them to the cloud. Until the data rate is relatively small, they can easily connect to cloud servers for data storage & analytics. But many medical devices used for continuous health monitoring produce data at a high rate and transferring that data into the cloud needs a high and uninterrupted internet connection. Edge computing solves this by analyzing the data and producing the report much faster and then periodically sending the computed data into the cloud.

Edge Computing is a boon for remote monitoring & care which helps healthcare services to reach all the places around the globe [10]. Those computed data can easily be accessed anywhere by the medical practitioner, and they can give their attention and care on time. In this way, edge technology reduces cost, triggers the alert for immediate attention on emergencies and safeguards the medical data in remote healthcare systems.

6.4 Blockchain

Blockchain technology is one of the emerging technologies that has grown in importance over the past decade. Blockchain is a robust solution for secure electronic communication. It was created together with cryptocurrencies, distributed ledgers and smart contracts. The application of blockchain technology has been analyzed by the research community and to the fame of the blockchain, it can be used in a wide range such as Bitcoin, Secure Data Storage, Telecommunications, Healthcare, IoT, Smart Cities, Cloud Computing, Resource Sharing and many more. Blockchain Technology is described in Fig. 6.7.

6.4.1 Transaction and block

Blockchain technology is built using basic cryptographic techniques. The strengths depend on the way the cryptographic techniques are applied to frame each of the blocks. Basic information that is stored in blocks is called transactions. A transaction is an application-dependent action that is collected in a block. Each transaction is hashed using a hashing algorithm, and the hash is then encrypted using the user's private key. This hash forms a digital signature for that transaction. The digital signature is sent to the network along with the data. The transaction is processed and validated by each validation node by decrypting the signature and comparing it with the hash generated by the validation node for this transaction. This validation process verifies the authenticity of the user who performed the transaction and the integrity of the data. The valid transaction is sent to the mining node to create the block.

As shown, blocks are containers of information or transactions. Blocks may contain any information such as a bank transaction, a picture, a program or the like. The validated information or transactions are packed into blocks. Each block records multiple transactions and secures these transactions based on the hash value of the previous block and the transactions accumulated in the block. As new transactions develop, new blocks are created and added to the previous block. The

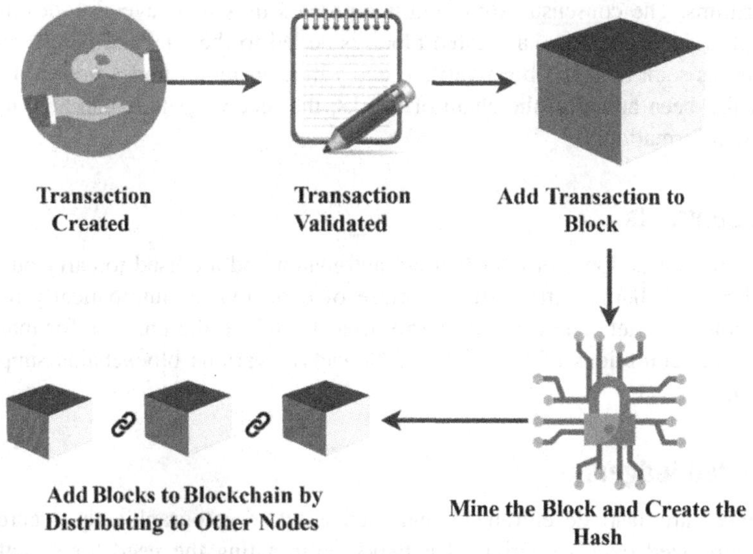

FIGURE 6.7

Blockchain technology.

first block, known as the Genesis block, has no relation to any previous block because it has no parent block. From the second block onwards, each block is linked to the previous block by hashing the block data along with the hash of the previous blocks, creating a chain of blocks known as the blockchain. Every block created is processed and validated by the majority of network participants, eliminating the need for a trusted third party. These blocks are stored in a distributed manner.

From an architectural point of view, blockchain is a linked list data structure based on Distributed Ledger Technology. All blocks are distributed to all machines connected to the network. Blocks are created by miners and protected by hashing algorithms. The mining node checks the valid transactions and groups them into a block so that the block size does not exceed a predetermined threshold. The hash of the previous blocks and the transactions of the current blocks are hashed together to create the hash for the current block. This form of hash creation forms the link in the blockchain and makes the block immutable. Manipulating a transaction in a block changes the hash value and all subsequent blocks. Updating the entire hash in the following blocks is difficult. This fulfils the goal of the blockchain to record the transactions in a distributed and immutable manner.

6.4.2 Consensus mechanism

A blockchain process is based on the consensus mechanism to verify a valid transaction. A blockchain network consists of various nodes that are distributed and connected in a decentralized manner. Every node that needs to execute a transaction needs to write the transaction to a block. All nodes present in the network have to agree on transactions, which are carried out based on consensus algorithms. The consensus algorithm is a process in which several nodes in a distributed network agree on a decision. Thus, a created block is added to the chain of blocks already present in the blockchain as soon as it has been verified and validated by all nodes present in the network. Once the block has been added to the chain of blocks, the receiving node can now update the ledger with the new information.

6.4.3 Smart contracts

Smart contracts are transaction protocols that are automated and are used to carry out the protocols mentioned in the blockchain contract. It is a piece of code that is automatically implemented if specified conditions are met. These are primarily used to reduce the chances for malicious exceptions and satisfy the conditions of the contract. Nowadays various blockchains support the smart contract paradigm.

6.4.4 Distributed ledgers

Distributed ledgers are fast, decentralized and, above all, cryptographically secured. It can be shared and synchronized over a distributed network, eliminating the need for a central authority. All information and records are cryptographically secured and can only be made accessible through the keys with cryptographic signatures. With the distributed ledger, all changes made in the ledger, or the document are transferred to all participating nodes.

6.4.4.1 Types of blockchain

Blockchains can be divided into public, private, consortium and hybrid blockchains.

1. A public blockchain is a blockchain setting that is publicly accessible without the need for access permission. The blockchain can be updated by any public.
2. A private blockchain is a blockchain setting that makes it private with restricted access. Such blockchains are controlled by a central authority.
3. A consortium blockchain is a shared blockchain where few organizations work together to build the blockchain.
4. The hybrid blockchain is a combination of the private and public blockchain. The viewing of transactions is made public, and the change is made private.

The selection of the blockchain type to be adopted depends on the application of the blockchain. The secure form of the blockchain is the public blockchain. As with other private and syndicated blockchains, there is a central authority to control the blockchain, which increases the possibility of manipulation.

Medical data privacy

The advancement of information technology has made it possible to collect health information from patients remotely through portable devices, and also to allow doctors to monitor patients remotely. This data, which is related to the health aspect of the patient, is known as an Electronic Health Record (EHR) or Electronic Medical Record (EMR) [11]. These data are sensitive personal data of the patient that must be available to the authenticated users, the doctors and the patients.

This would enable diagnosis at an early stage, remotely and at a low cost. The medical records are extensive data that are very useful for improving treatment by continuously tracking the health status and treatments performed, predicting disease, the cause of disease and possible preventive measures [12]. All of this is made possible by the data mining systems. The data mining system has to be trained, for which it requires a large amount of data.

Medical data storage and sharing

Medical data collected from patients must be captured, securely stored and made available to doctors without compromising security, the data is saved as a hard copy or on a storage device, it can be corrupted and there is no way to retrieve the data. The data, when stored locally, requires a separate data-sharing infrastructure for remote access to the data. The security of the stored data also depends on the security of the firewall, access control mechanisms, virtual private network (VPN), etc. used in the hospital management system. There is a potential for data loss due to tampering that can affect the reliability, integrity, availability, and confidentiality of the data. In addition, the patient data from each hospital is stored separately in its database. It becomes difficult for patients to get their health information and share it with different doctors in different hospitals. On the other hand, the data stored in the cloud increases the availability of data, but there is a possibility of a data breach. The cloud also follows a centralized management system in which a single party must be trusted for the security of the stored data. The other problems with centralized data storage systems are that the same patient data is available in multiple hospitals and that data is not readily available to the patients themselves.

Security breach

A security breach on healthcare data is very much possible due to the insecure storage in a centralized system and sharing the data without access restriction and verification. Table 6.2 shows the number of health records compromised for the past year [13]. Table 6.3 lists the number of data breaches of above 500 records for the past 10 years [14]. From the table, it is seen that the amount of healthcare data breaches is increasing every year. Various causes for the compromise are Ransom attacks, Hacking, Online exposure of data, Insider attacks, and Phishing. Hence it is necessary to protect this information from tampering while storing it in a distributed system that would increase the availability of data.

6.4.4.2 Blockchain for secure storage and sharing in the healthcare management system

Blockchain technology is a boon to the health management system. It is a tamper-proof and secure technology that enables safe medical data storage and sharing. Since the blockchain is based on distributed ledger technology, it increases the availability of data. In recent years, the blockchain has received immense attention in healthcare as it provides a way to secure healthcare data and also maintain its privacy in a distributed and trustless environment. It increases the interoperability, scalability, reliability, integrity, availability, and confidentiality of the data facilitated by the distributed ledger and blocks principles in blockchain technology.

Table 6.2 Number of health records compromised for the past one year.

Month	November 2020	December 2020	January 2021	February 2021	March 2021	April 2021	May 2021	June 2021	July 2021	August 2021	September 2021	October 2021
Number of health care records compromised (in thousands)	1139	4241	4467	1234	2913	2583	6535	1290	5570	5120	1253	3589

Table 6.3 Data breach for the past 10 years.

Year	2011	2021	2013	2014	2015	2016	2017	2018	2019	2020
Data breach of 500 or more records	199	219	277	314	270	329	358	368	512	642

6.4.4.3 Digital IDs

According to a Health Insurance Portability and Accountability Act (HIPAA) survey, the average healthcare worker has access to 31,000 sensitive health records on their first day of work. The healthcare industry has the highest data breach cost, which is $7.13 million per data breach, according to the IBM Security Cost of a Data Breach Report, 2020. The main causes of security breaches are poor access control systems and the lack of proper verification systems. Every user, including the patient, physician, and health care professional, must be verified and given restricted access.

Digital identification—DID

Identification is very much essential for one to prove himself. Digital identification (ID) plays a crucial role in authenticating and verifying a person. If used in the healthcare sector it can provide efficient health care to a patient. It eliminates the need to carry all the health documents whenever a patient meets a healthcare professional. It would help the healthcare professionals to have a complete view of the patient's health record and treatment history, which would improve the quality of the patient's treatment and retention. Blockchain-based healthcare systems need to use DID to check and restrict access.

Content identification—CID

Blockchain technology is not suitable for storing files of huge sizes. It is an expensive medium for storing medium to large size files. Imaging data such as MRIs creates files that are often more than 200 megabytes in size. This makes using the blockchain to store these files impossible due to the inherently high network latency. To store such data and have the same security and availability as blockchain technology, it is better to store the file encrypted in the storage location of the health center or the cloud. In Ref. [15], the authors proposed a blockchain-based searchable encryption scheme where each EHR is identified by its unique Content ID (CID) which forms an index to the EHR. The CID is constructed through complex logical expressions. The CID index will form a transaction in the blockchain. This CID along with the hash of the block is stored in the block. The CID index can be used to access the EHR stored in the cloud. The data owners in addition will have full control over their data. In Ref. [16], the authors have proposed a similar scheme where a proxy re-encryption technique is used to provide access control to the cloud-stored EHR and the related metadata are stored on a private blockchain.

6.4.4.4 Distributed immutable patient records

Blockchain is a decentralized immutable chain of records. Blockchain can be built using a distributed ledger where the data once stored cannot be tampered with and can be shared. This tamper-proof is made possible through the chain of blocks and their link through the hash value recorded in each of the blocks. This feature makes blockchain more suitable for maintaining the patient's record in an immutable manner.

In Ref. [12], the authors have proposed a Hyperledger-based electronic healthcare record (EHR) sharing system. Their proposed system architecture and algorithm follow a patient-centered approach to securing and sharing the data. It provides an access control policy with symmetric keys to different healthcare providers which would increase the confidentiality of the patient's data. A blockchain-based Electronic Rehabilitation Medical Record (ERMR) sharing scheme has been built [17]. The design is based on specific technologies of blockchain such as hybrid P2P network, blockchain data structure, asymmetric encryption algorithm, digital signature and Raft consensus algorithm.

6.5 Medical imaging modalities with the Internet of Things

Internet of Things (IoT) becomes an inevitable technology that extends internet connectivity across various devices and objects used in day-to-day activities. It offers a wide range of distributed environments to connect smart devices in smart cities by smart people. Most of the devices in the hospitals and homes are connected through IoT to monitor, control and share data with healthcare providers and users. Such devices provide seamless integration of data at the level of patient information, data collection, sharing, monitoring, billing, scheduling, and physician consultation for optimal productivity. A similar technological approach is being extended with imaging modalities for order entry, effective sharing of images acquired, assessment/review of images, real-time consultation, and other imaging services. It can also be extended for diagnosis, troubleshooting and calibration of imaging modalities by the service engineers remotely.

The idea behind interconnecting and monitoring medical imaging equipment over the internet originated 20 years back. Three major medical imaging giants Philips, Siemens, and GE (general electric) came up with an "All-in-one-cloud" project to march towards the Health cloud ecosystem rather than computer intense image processing technique. Building a self-sufficient health ecosystem and increased cost-effectiveness are the benefits of the cloud-based medical imaging system. Next to genomics, medical imaging is the fastest growing and essential sector in healthcare support for disease prediction, diagnosis, prognosis and surgical planning. The amount of medical data being generated by a modality in a hospital has become tripled in the year 2010. Terabytes of data need to be stored in an organized manner and to be handled in an effective manner using the appropriate platform. Web-based imaging analytics platforms with supercomputing infrastructure ease the computational process thereby processing time can be reduced with minimal human effort.

6.5.1 Machine to machine communication

Machine to machine communication (M2M) is a direct method of data sharing between any two devices without any human involvement by communicating through the wired or wireless mode. This particular approach has been very successful in industrial automation. In the PACS environment, the involvement of radiologists, technical assistants, and data entry operator's role is laborious. The annotation and reviewing of medical images take extensive time and effort by the professionals to store and share the data. Even the data being transferred from image acquisition to the gateway and from PACS server to workstation needs human involvement in all aspects. Natural language processing, automated annotation using object tracking, and voice over technologies can be adopted in the PACS ecosystem to simplify the data labeling, description, and representation.

6.5.1.1 Smart imaging devices

Regarding image interpretation, the invention of smart electronic devices such as Ipad, and mobile devices have many advantages to observing the images for faster diagnosis and treatment in remote places [18]. Massachusetts general hospital developed a project to utilize these smart devices for Tele-stroke studies. Without much storage, the images can be viewed immediately by the doctors. Thereby superior inter-observer agreement in the identification of critical findings is easier than the earlier approaches. Multiple applications available in the smart devices support the clinicians to review the images from different perspectives during rounds and in the operation room. Such devices act as mobile PACS workstations for the doctor with all additional features such as image denoising, image enhancement, and essential tools for cropping, measuring and visualization as shown in Fig. 6.8.

6.5.1.2 Mobile imaging devices

Recently, mobile phone technology in the radiology department has introduced many innovations as shown in Fig. 6.9. Those products are portable, handy, doctor-friendly, inexpensive, and have more accuracy than conventional portable devices. Mobisante has introduced a mobile-based

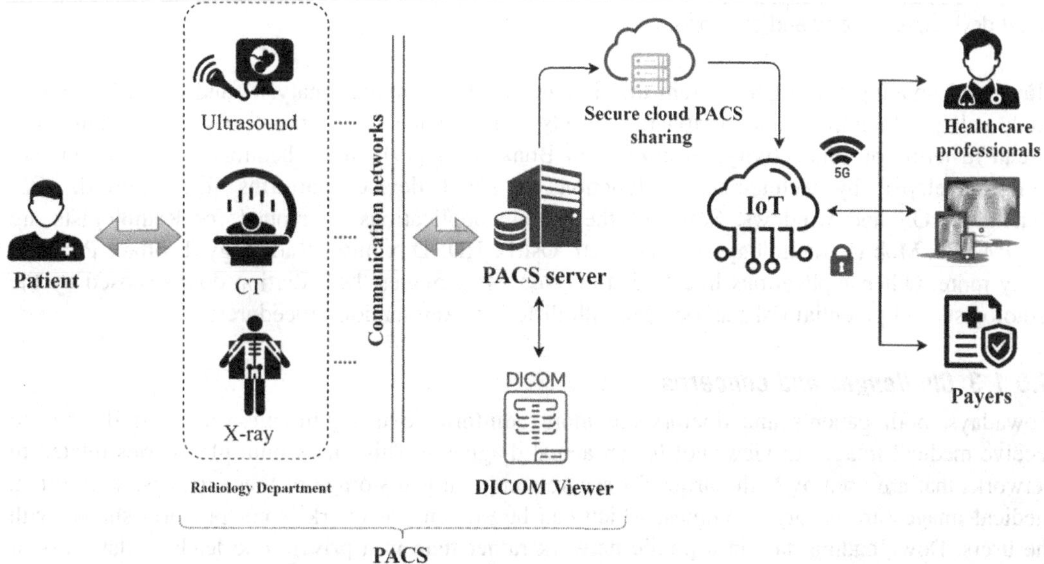

FIGURE 6.8

Internet of Medical Imaging Things (IoMIT).

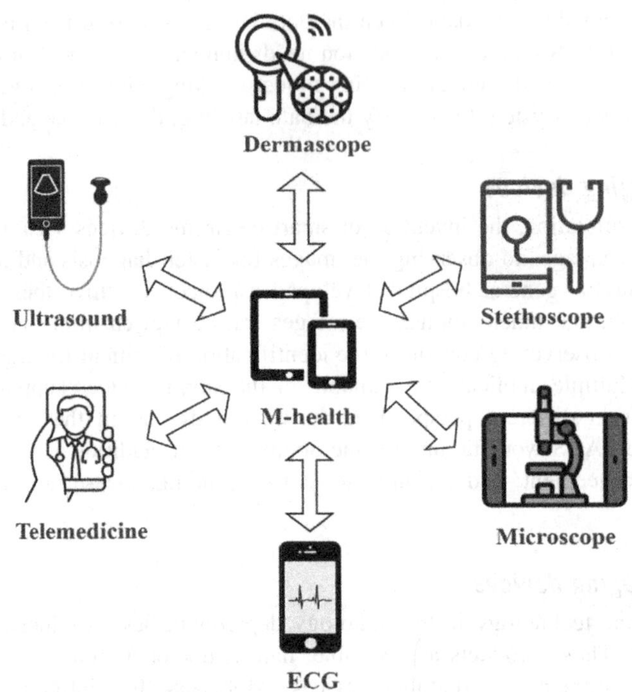

FIGURE 6.9

Smart devices for imaging and diagnosis.

ultrasound imaging (MobiUS) system that is intended for imaging, analysis, and measurements in fetal, pelvic, abdominal, and peripheral vessels. This device is comparatively better than GE's Vscan in terms of connectivity Wojtczak and Bonadonna [19]. Many healthcare applications are being developed by technocrats to function on smart device platforms like Android, iOS, Blackberry OS and windows. Some of the mobile applications developed for Radiologists are LifeIMAGE Mobile, Radiology Assistant 2.0, Osirix HD, Doximity, Radiology Toolbox Pro, and many more. Other applications like Radiation passport provides the radiation dose exposed by the radiologists and potential risks associated with different examination procedures.

6.5.1.3 Challenges and concerns

Nowadays, both patients and doctors are more comfortable using mobiles or smart devices to receive medical images to view, not for an actual diagnosis. This raises multiple queries related to networks that are used by both parties for secure reception and storage of these images. Very often, medical images are in large volumes, which can bog down a network if not properly shared with the users. Downloading data in a public network rather than in a private one leads to data loss or theft. Food and Drug Administration (FDA), USA is more concerned about authorizing mobile based devices that are displayed accurately, image content without loss of contrast details and can accommodate a wide viewing area.

6.5.1.4 DICOM matters

The mobile devices utilized for viewing medical images should meet DICOM—compliance, which means maintaining the image quality similar to high-resolution imaging workstation displays. A DICOM calibrated imaging display provides precise image representation to the radiologists who can diagnose the images from different modalities like CT, MRI, Ultrasound, PET etc. A calibrated tablet can be utilized to view large images with actual information when the workstation display is not available. Such tablets can also be shared with the peer members when consultation is required.

6.5.2 5G on the Internet of medical imaging things

5G is the fifth-generation cellular network that significantly helps to improve the network connection by bringing new features that enable to delivery of solutions across society. In a 5G network, machine to machine communication and IoT play a vital role in the healthcare system. To support smart healthcare IoT-based app technologies, 5G networks are designed to be scalable and robust, needing not only a high data rate but also other features like large convergence, dense distribution, stability, low latency, high energy efficiency and long-range communication. Body area network involves short-range wireless technologies like Bluetooth, Zigbee, 4G, radio frequency identification (RFID), and Wi-Fi whereas a smart health system demands technology that facilitates the large data transfer between the local server and base station. The various IoMT applications of communication standards are listed in Table 6.4.

The technology should also meet other needs like battery life, device simplicity, and interoperability among various service providers. Smart healthcare systems can be classified as Things-oriented, App-oriented, and Semantics-oriented. Deployment of 5G enabled smart home—health

Table 6.4 Wireless field mediums and their application in IoMT

Standards	Range (m)	Data rate	Frequency	Applications
Zigbee	10–20 m	250 kb/s	868/ 915 MHz; 2.4 GHz	Bowel activity, monitoring device, Digital stethoscope, Heart rate monitor
Bluetooth	10–100 m	1 Mb/s	2.4–2.5 GHz	Digital stethoscope, fit bit tracker,
RFID	Up to 3 m	Varies with frequencies	LF/HF/UHF/ Microwave	Medical equipment tracking system, the wearable RFID tag antenna
Near Field Communication (NFC)	<0.2 m	Up to 424 Kb/s	13.56 MHz	Alzheimer's support center, Implanted data acquisition systems, home monitoring,
Wi-Fi	45–90 m	Up to 54 Mb/s	900 MHz to 60 GHz	Wireless health monitoring, indoor localization,
3G and 4G	1–8 km	Varies on the network type	Varies on the network type	Wireless health monitor
Satellite	160–3600 km	Varies on model	Varies on model	Disaster conditions, Real-time wireless health monitors

care requires enhanced mobile broadband, massive machine-type communication, low latency, high-reliable communications, and a wireless regional area network. App-oriented architectures ensure the authenticity of data transfer between apps on smartphones and sensors. [20] discussed the flexibility, low power consumption, high efficiency and switching on the intelligent procedures that can be achieved through things-oriented architecture whereas natural language processing, prediction and analysis can be achieved through semantic oriented architecture.

6.5.3 Blockchain and edge computing in IoMIT

As discussed previously (in Sections 6.5.1 and 6.5.2), the various imaging modalities available in the super specialty hospitals or any network hospitals can be connected to the IoT environment to establish a link among the imaging and non-imaging devices to share the resources and data at the instance.

The conventional method of storing and retrieving the files is from distributed databases or cloud-based databases. Such transactions lack the performance of the PACS whenever the data is to be maintained confidential. Similarly, the image retrieval approach happens in three different scenarios: (1) Image intelligent retrieval from multiple sources (2) Image retrieval for privacy protection concerns (3) Image retrieval for tracing continuous analysis. In those conditions, the data being generated and stored in the stand-alone database or Cloud-based ecosystem would not help the users or administrators to access the data immediately.

The images that are acquired from various stationary and medical imaging devices are mined and shaped in different blocks as shown in Fig. 6.10. The blocks may have information about the patient, image orientation, imaging modality, DICOM information, and reports generated by the administrators. All the blocks can be secured by cryptographic technique. Depending on the accessibility, the blocks can be private, public, consortium or hybrid as explained in Section 6.4.1. The Digital IDs can also be linked with the blocks if the data is expected by the national agencies and by the health insurance payers. The Blocks containing information are stored in the edge ecosystem to access the data based on the need. To proceed with high dimensional image analytics, image modeling, and retrieving similar images in 2D and 3D of the multimodal dataset are possible with this proposed framework. Each PACS are considered a node that is connected through the high-speed data communication medium 5G. Edge computing enables direct machine–machine communication to provide optimal results to the end-user without any human intervention. Artificial Intelligent systems embedded in the edge layer can introduce supervised, unsupervised and reinforcement learning to the IoMIT ecosystem thereby medical image classification, retrieval and image modeling are possible for effective surgical planning and procedures.

As discussed in Section 6.1, retrieving medical images can be done by any of the following approaches: TBIR, CBIR, SBIR and hybrid. In the traditional retrieval approaches, patient information is not considered with image feature data. Moreover, the security concern is the least bothered in all the approaches. Considering these issues, the image retrieval service receives the encrypted features of all images (both training and query) on the blockchain by accessing the public address and storing them in a local database. All image information is indexed to improve the overall performance of the system. Usually, retrieval system performance is estimated by precision, recall, and F-score. The system updates the image indexing very often whenever the data flows in. If a third party likes to access the data, the miner needs to grant temporary access to them through a smart contract. The training images and their features are properly indexed, encrypted and stored in the cloud environment on the blockchain. If the user sends a query request to the retrieval system,

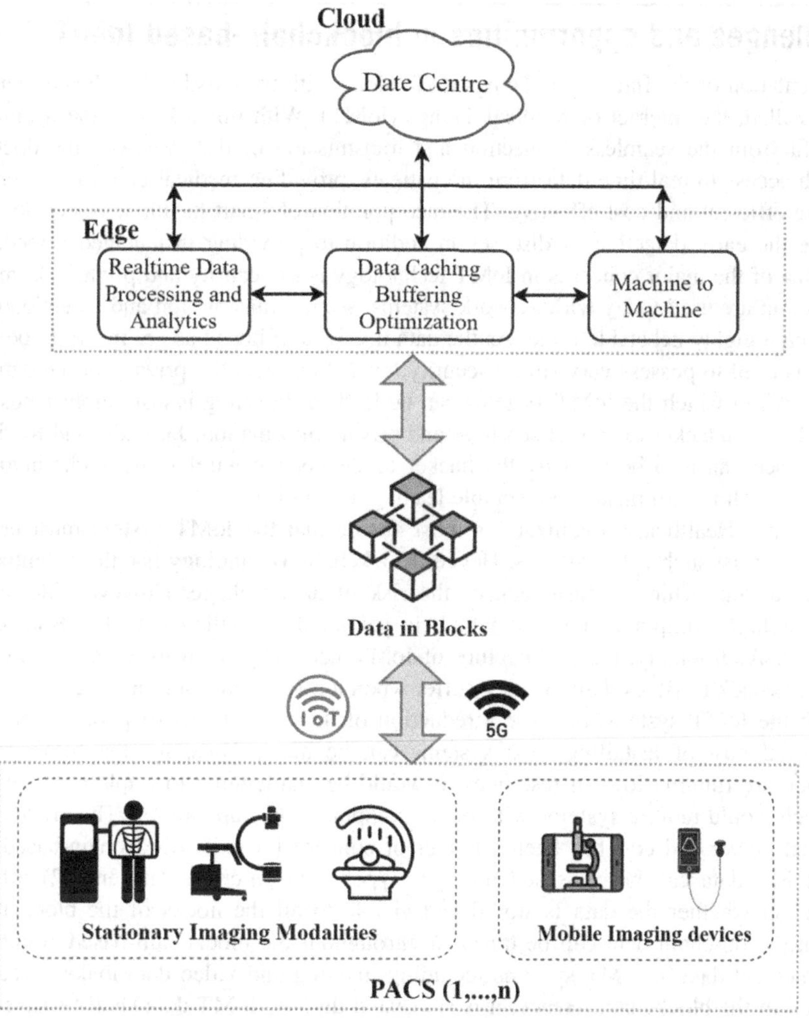

FIGURE 6.10

Layered architecture of IoMIT using Blockchain-enabled Edge computing.

the query features are compared with the stored features at the edge level. The relevant information and retrieved images are provided to the user. Generally, Euclidean distance is successful in comparing the feature datasets of training and testing images. To deal with huge data, growing day by day, features can not be confined to multi-modality images. To skip feature engineering (feature extraction and feature selection), traditional machine learning algorithms may not be useful. Hence, the best alternate solution is to deploy deep learning-based architectures in this framework. Recently, deep learning models are widely used for various applications to deal with high dimensional data with less complexity and less human intervention.

6.6 Challenges and opportunities in blockchain-based IoMIT

The implementation of the Internet of Things (IoT) along with the existing healthcare facilities led to a new term called, the Internet of Medical Things (IoMT). With this addition, the healthcare sector would benefit from the seamless connection and transmission of data between the doctor and the patient. With access to real-time data from the patients, providing medical care to the concerned has become more efficient and cost-effective. The incorporation of smart health sensors into devices can also promote the early detection of diseases in addition to providing instantaneous medical data to clinicians. One of the major concerns in IoMT technology is its security and privacy. In most modern technologies that are used today with network systems, secure transmission and reception of information have been a highly debatable topic. As the data involved in hospitals are meant to be private, the IoMT systems need to possess very strong security levels to protect the privacy of the patients. There are several ways in which the IoMT systems can be hacked. Phishing is also another possibility that can be used by the attacker to extract sensitive and private information. Jammers and Radiofrequency (RF) Interference can also be used by the hacker to disrupt the wireless networks involved in the IoMT system which in turn make it vulnerable leading to data leak.

In this regard, Healthcare organizations must ensure that the IoMT system must be extremely secure and can resist such cyber-attacks. Hence, Blockchain Technology has the potential to handle such security attacks which in turn reduces the risk of data leakage. However, blockchain algorithms have a high computational cost which is yet another challenge to the Blockchain-based IoMT systems. Additionally, the architecture of IoMT networks has a lot of nodes involved in it, which is a drawback to Blockchain as they suffer when a lot of nodes are involved.

Although the IoMT systems promise a reduction of the overall cost of patient care in the long term, the initial cost of installing such systems can be huge. Thus, for the IoMT systems and devices which are running low on resources, it would be challenging to implement the blockchain systems which would require systems with better computational capabilities. This trade-off between computational power and cost has been a matter of concern for such Blockchain-based IoMT systems. The IoMT data can be classified into two types—(1) on-chain data and (2) off-chain data depending upon whether the data is stored and visible to all the nodes of the blockchain or not. Even though the on-chain data can be traceable throughout the blockchain-based system, the huge amount of medical data in IoMT systems including imaging and video data make it challenging to store them all in the blockchain. Hence, it is suggested that the IoMT data shall be saved in an off-chain method and the blockchain may store the metadata of the off-chain IoMT data. This technique can save huge storage costs for Blockchain-based IoMT systems.

With the integration of blockchain networks to IoMT systems, there is a scope for improvement in the security aspect of these systems. The built-in encryption and digital signature features along with authentication can supplement the system's security. As the amount of medical data in healthcare organizations is increasing exponentially, there is a need for Big Data systems to manage this huge data. The blockchain-based IoMT systems can be enhanced with such big data analytics which can be used to analyze both IoMT and blockchain data. Going forward, cloud-based systems would be required to store and process such large data in the IoMT systems. This dependency on third-party storage brings out a risk of data leakage and privacy concerns. The integration of blockchain networks to such cloud-based systems can efficiently secure the IoMT data. The recent advancements in Artificial Intelligence and Deep Learning techniques are also promising for their future integration into the Blockchain-based IoMT systems.

6.7 Summary and conclusion

With an increase in the growth of several advanced technologies at a rapid pace, there arises an opportunity to incorporate such state-of-the-art methodologies into the medical institutions to enable the delivery of efficient, accurate and ubiquitous healthcare to patients. One such area of interest to researchers is the management of the large medical records databases including EHR, imaging data and lab results of the patients in the healthcare setup. In this study, we propose a secure system for the storage and retrieval of medical images in a hospital database with the Internet of Medical Things (IoMT) by integrating it with both Blockchain and Edge Computing techniques for added security and privacy. As the information in databases of the healthcare organizations is usually sensitive data such as the patient EHR and medical imaging data, there is an utmost need to ensure its privacy. This has been addressed with the fusion of blockchain-based IoMT systems along with Edge Computing that proves to be secure and reliable to prevent any data breach in such large and confidential databases. Such blockchain-enabled storage of medical images in a hospital's database provides a decentralized data management server that can store and retrieve medical images accessed by several users on the server.

Blockchain provides several benefits in terms of speed and reliability to the end-user. It eliminates the use of a single centralized system that causes a single point failure in the system. The usage of blocks to store medical information provides a streamlined workflow for the storage and retrieval of patient data. Edge Computing is another latest technology that is used along with IoMT to store large amounts of medical data. This has overtaken the existing Cloud Computing systems which face huge latency issues when faced with extreme amounts of data coupled with low telecommunication bandwidths. In Edge Computing, the data is stored and retrieved at a site that is very close to the original data source instead of performing the computational processes at a remote site. It is usually performed at the hospitals, clinics or even directly on the patient monitoring devices. The most promising benefit of this technique is its low time delay. It pushes for a quick and real-time application of medical image storage and retrieval.

The proposed architecture can also be extended to medical image classification with the incorporation of feature engineering techniques such as Deep Learning and Image Processing. This can aid in faster diagnosis and early detection of diseases. The study presented in this work may serve as a premise or inspiration for future work-oriented in this direction of secure Blockchain-enabled Edge Computing for IoMT. Thus, with the fulfilment of several challenges including the secure storage, user-centric nature, inter-workability and efficiency with low latency, Blockchain systems along with the Edge Computing techniques can be a game-changing technology for the hospitals and clinics to manage such large medical databases with a user-friendly interface to the doctors and clinicians.

References

[1] D. Young, D.S. Young, K.N. Chul, Image retrieval using BDIP and BVLC moments, IEEE Trans. Circuits Syst. Video Technol. 13 (9) (2003) 951–957.

[2] H.H. Khaleel, R.O. Rahmat, D.M. Zamrin, Components and implementation of a picture archiving and communication system in a prototype application, Rep. Med. Imaging 12 (2018) 1–8.

[3] S. Dash, S.K. Shakyawar, M. Sharma, S. Kaushik, Big data in healthcare: management, analysis and future prospects, J. Big Data 6 (1) (2019) 1–25.
[4] V. Jeyakumar, B. Kanagaraj, A medical image retrieval system in PACS environment for clinical decision making, Intelligent Data Analysis for Biomedical Applications, Academic Press, 2019, pp. 121–146.
[5] M. Alkhawlani, M. Elmogy, H. El Bakry, Text-based, content-based, and semantic-based image retrievals: a survey, Int. J. Comput. Inf. Technol. 4 (01) (2015) 58–66.
[6] A. Qayyum, S.M. Anwar, M. Awais, M. Majid, Medical image retrieval using deep convolutional neural network, Neurocomputing 266 (2017) 8–20.
[7] P. Kurian, V. Jeyakumar, Multimodality medical image retrieval using convolutional neural network, Deep Learning Techniques for Biomedical and Health Informatics, Academic Press, 2020, pp. 53–95.
[8] F. Casino, T.K. Dasaklis, C. Patsakis, A systematic literature review of blockchain-based applications: current status, classification and open issues, Telemat. Inform. 36 (2019) 55–81.
[9] G.J. Joyia, R.M. Liaqat, A. Farooq, S. Rehman, Internet of medical things (IoMT): applications, benefits and future challenges in healthcare domain, J. Commun. 12 (4) (2017) 240–247.
[10] V. Jagadeeswari, V. Subramaniyaswamy, R. Logesh, V. Vijayakumar, A study on medical Internet of Things and Big Data in personalized healthcare system, Health Inf. Sci. Syst. 6 (1) (2018) 1–20.
[11] S. Shi, D. He, L. Li, N. Kumar, M.K. Khan, K.K.R. Choo, Applications of blockchain in ensuring the security and privacy of electronic health record systems: a survey, Comput. Secur. (2020) 101966.
[12] S. Tanwar, K. Parekh, R. Evans, Blockchain-based electronic healthcare record system for healthcare 4.0 applications, J. Inf. Secur. Appl. 50 (2020) 102407.
[13] Steve Alder, 2021. Available at: https://www.hipaajournal.com/october-2021-healthcare-data-breach-report/
[14] Health Insurance Portability and Accountability Act, 2020. Available at: https://www.hipaajournal.com/healthcare-data-breach-statistics/
[15] L. Chen, W.K. Lee, C.C. Chang, K.K.R. Choo, N. Zhang, Blockchain-based searchable encryption for electronic health record sharing, Future Gener. Computer Syst. 95 (2019) 420–429.
[16] T.T. Thwin, S. Vasupongayya, Blockchain-based access control model to preserve privacy for personal health record systems, Secur. Commun. Netw. 2019 (2019).
[17] J. Zhang, Z. Li, R. Tan, C. Liu, Design and application of electronic rehabilitation medical record (ERMR) sharing scheme based on blockchain technology, BioMed. Res. Int. 2021 (2021).
[18] S. Gupta, E.M. Johnson, J.G. Peacock, L. Jiang, M.P. McBee, M.B. Sneider, et al., Radiology, mobile devices, and Internet of Things (IoT, J. Digital Imaging 33 (3) (2020) 735–746. Available from: https://doi.org/10.1007/s10278-019-00311-2.
[19] J. Wojtczak, P. Bonadonna, Pocket mobile smartphone system for the point-of-care submandibular ultrasonography, Am. J. Emerg. Med. 31 (3) (2013) 573–577. Available from: https://doi.org/10.1016/j.ajem.2012.09.013.
[20] L.Z. Yue Li, X. Wang, Flexible and wearable healthcare sensors for visual reality health-monitoring, Virtual Real. Intell. Hardw. 1 (4) (2019) 411–427.

CHAPTER

Lane detection and path prediction in autonomous vehicle using deep learning

7

Renu Kachhoria[1], Swati Jaiswal[2], Meghana Lokhande[1] and Jay Rodge[3]

[1]Department of Computer Engineering, Pimpri Chinchwad College of Engineering, Pune, Maharashtra, India

[2]SCOPE, VIT University, Vellore, Tamil Nadu, India [3]NVIDIA AI, San Francisco, CA, United States

7.1 Introduction

In recent years, providing appropriate traffic safety has become a concern among research professionals, organizations, and government affiliates. As per the details available from the US Division of Transportation, 36,560 people were reported dead due to motor vehicle-related mishaps in 2018, inferring that there were, on average, 100 deaths every day due to vehicle accidents. It was observed that 94%–96% (as per the survey) of motor vehicle crashes were due to human error. In the current scenario, many advancements have been made in the field of driver help assistance systems, such as device stability control and lane prediction/detection. Lane departure warning also fosters security and abatement of driver load, providing the motivation for automated driving technologies.

The Society of Automotive Engineers (SAE), a US-based professional association for engineering professionals, has portrayed six particular levels of driver help development movements. Level 0 attributes can simply provide advice and temporary assistance, e.g., providing blind-spot warnings, applying emergency breaks, etc. Level 1 and 2 attributes can give guiding or/and brake/speed increase control to the driver. For Level 3 or higher frameworks, the drivers are not driving when these technology features are occupied. Level 3 independent driving elements require the driver to dominate and drive when the features are required. Furthermore, Level 4 elements can drive the vehicle under restricted situations, such as on the thruway, without drivers taking over. At long last, Level 5 elements can drive the vehicle fully automated i.e., in all situations. More elevated-level independent driving innovation requires better climate recognition capacities to distinguish the encompassing motors and their related expected risks involved.

Lane detection is one of the important challenges for autonomous vehicles, which has drawn the consideration of the computer vision local area for quite a long time. Basically, path discovery is a multifeature recognition issue that has become a real challenge for computer vision and AI methods. During the last few decades, a couple of functionalities toward automated driving have been incorporated into vehicles, such as monocular vision and sound framework vision modes for camera-based vision structures. In front-view-based applications, such as adaptive cruise control

Intelligent Edge Computing for Cyber Physical Applications. DOI: https://doi.org/10.1016/B978-0-323-99412-5.00012-5

(ACC) systems, monocular cameras achieve attractive precision over relevant distances with the standard point of convergence.

7.1.1 Previous work-related analysis

Over the last many years, multiple tools, sensors, and other functionalities have been introduced in vehicles to make them more adaptive and useful for autonomous driving. To achieve success in the adoption rate of autonomous driving various researchers and industries are gradually incorporating multiple technologies and services in the field.

1. The acknowledgment of path change of a off track driving vehicle is important to work on the intellectual capability of independent features driving frameworks [1]. In this review, an extensive identification model was used to decide the lane change status of a main vehicle, and this was done by breaking down the sidelong distance and sidelong speed of the main vehicle by using millimeter-wave radar information obtained in a naturalistic driving test. The experimental results showed that the acknowledgment precision pace of the BPNN (back propagation neural network) model expanded from 80% to 87% after PSO (particle swarm optimization) improvement at TLCs (time-to-lane crossings) longer than 1.0 s. The consequences of this review demonstrate that the execution of the proposed complete recognizable proof model for distinguishing the path change status of driving vehicles altogether expanded the ID exactness, consequently giving a premise to work on the presentation of intelligent insight frameworks. As of now, the proposed model is appropriate for path change which can distinguish straight paths and bend paths.
2. The research by Wang et al. [2] intends to introduce the researchers to a novel technique of path location and path following, which is a practical prerequisite for implementing driving assistance systems (DAS), such as lane take-off warning and lane keeping assist. To accomplish its proper outcome, the computerized picture handling was separated into three different levels. At the lowest level, the data image dimensionality is diminished from three layers to one layer, sharpness is improvised, and vicinity is characterized based on the minimum safe distance maintained from the vehicles ahead. The component extractor for path edges recognition configuration is essential for the intermediary level handling. The improvement in the path following technique is examined at significant levels. Hough transform and a shape-protecting spline insertion are used to accomplish a lane detection level. The trial results were subjectively and quantitatively assessed using a real observation examination. The system shows great precision levels, incorporating situations like shadows, images, bends, and street slant.
3. In Feng et al. [3], from subjective examination and results of calculations, it was possible to identify situations where the calculations are followed more decisively, along with highway pattern designs or ecological conditions where the prediction of the path is not stable. It was seen that the deceptiveness of the lane markings could be undermined by a few variables, such as the reflection of the lane, camera's glare, shadow images, and the crumbling of the paint. The most noteworthy decisiveness records were seen in following the uninterrupted lane markings; the shallow performance indexes were noticed double marked with the internal lane dashed lines.
4. In their paper, Zou et al. [4] propose a method using high-resolution automotive radar for detecting lane markings in which radar sensors distinguish the road boundaries on the basis of

objects like guardrails, delineators, street checks, etc. Additionally, in a multipath street or with a missing side of the road foundation, the neighboring paths and the street limit must be perceived by the radar sensors. It is important to fit radar reflectors into the path or street limits. The reflectors should be placed low on the streets so that it does not damage the vehicle tires. However, they must have the option to mirror the radar signals appropriately. The dispersing property of the different kinds of reflectors is assessed by simulations over a wide angular range geometry. The simulation results regarding the viewpoint effect on the radar cross area are investigated. In general, this paper demonstrates how to achieve a path recognition technique in reality using a high-frequency auto radar sensor in combination with road reflectors for an ideal radar reflectivity.

5. An advanced technique proposed by Luo et al. [5] consists of a back-propagation (BP) neural network model enhanced by incorporating a particle swarm optimization (PSO) algorithm. A persistent identification model is created based on real-time tests utilizing millimeter-wave radar information. The PSO-BP neural network model is prepared using actual vehicle lane change information and carried out when the time-to-lane crossings (TLCs) of the main vehicle is longer than 1.0 s. In comparison with the BP neural network model, the acknowledgment accuracy of the advanced model increases from 80% to 87% after PSO advancement with a window of period 1.0s; these outcomes meet the acknowledgment prerequisites of the independent driving frameworks for far off targets.
6. To this end, various path locations using different edges of persistent driving scenes was examined, and proposed a new hybrid deep learning model by combining two strong methods, i.e., convolutional neural network (CNN) and recurrent neural network (RNN) by Lu et al. [6]. In particular, data of each casing is disconnected by a CNN block, and the CNN features of numerous persistent edges, consisting of the property of time-series, are then taken care of into the RNN block to include learning and path prediction.
7. In He et al. [7], the authors propose a novel and powerful multiple lane-detection algorithms based on and about structure data, which contains five integral limitations: length imperative, equal requirement, conveyance imperative, pair requirement, and uniform width requirement. Each of the five imperatives is joined into a Hough transformation based brought-together structure to choose path applicants. Almost close to 100% of the false alert applicants in the Hough transformation space can be eliminated.
8. Gurghian et al. propose [8], a novel lane detection framework, called scene understanding physics-enhanced real-time (SUPER) algorithm. The proposed technique comprises two principal modules: (1) a multi-leveled semantic division network as the scene feature extractor and (2) a physical science upgraded multipath boundary streamlining module for lane inference. The proposed framework was trained on utilizing heterogeneous information from Cityscapes, Vistas, and Apollo, and the exhibition was assessed on four separate datasets (that were never seen), including TuSimple, Caltech, URBAN KITTI-ROAD, and Mcity-3000 Table 7.1 illustrates the comparative study of various methods used for lane detection and path prediction.

7.1.2 Challenges

Lane detection is an important technology required in numerous applications, including navigation [9,10], driverless vehicles [11,12], departure alerts [13], and traffic engineering. The efficiency of

Table 7.1 Comparative study of methodologies used for lane detection and path prediction.

References	Title of paper	Methodology used	Pros	Cons
Andrade et al. [1]	A Novel Strategy for Road Lane Detection and Tracking Based on a Vehicle's Forward Monocular Camera	Hough transformation, interpolation	Better lane departure warning and lane keeping assist; good accuracy levels, including scenarios with shadows, curves, and road slopes.	Takes more time to process. Not suitable for night mode
Wang et al. [2]	Cognitive Competence Improvement for Autonomous Vehicles: A Lane Change Identification Model for Distant Preceding Vehicles	Back propagation neural network and particle swarm optimization	Proposed a compiled model using back propagation neural network and optimized by particle swarm optimization followed by a continuous identification model of time-to-lane crossing.	Time-to-lane crossing was carried out to address drawbacks of particle swarm optimization back propagation neural network.
Feng et al. [3]	Lane Detection with a High-Resolution Automotive Radar by Introducing a New Type of Road Marking	Hough transformation	Implemented methods using radar instead of a camera against situations like fog, gaze, glare, rain, etc.	More pattern extraction can be demonstrated.
Zou et al. [4]	Robust Lane Detection from Continuous Driving Scenes Using Deep Neural Networks	Hybrid deep architecture combined with CNN & RNN	Proposed methodologies to handle difficult situations in lane detection by using SegNet-ConvLSTM	A longer sequence of inputs was used to make improvements on the performance.
Luo et al. [5]	Multiple Lane Detection via Combining Complementary Structural Constraints	Hough transformation, dynamic programming	Proposed methodologies to remove impossible candidates, followed by optimization of remaining candidates.	Dynamic programming can be improved to add more features.

lane change monitoring and prevention has a significant impact on the comfort and security of autonomous vehicles. In most cases, intelligent vehicle systems are able to keep a safe distance from the leading vehicle without the driver's interference. A problem arises when another automobile from the adjacent lanes begins cutting near the autonomous vehicle. If the driver is unable to take control, the autonomous vehicle will most likely collide with the cut-in vehicle. An early and smooth reaction is possible when such situations are recognized in time. It is also important for the autonomous vehicle to predict lane change intent when it is trying to change lanes. To function properly, the autonomous driving system must ensure that no other cars are changing lanes or moving to a certain distance in the indicated direction. If a system identifies such a vehicle, it should promptly cancel changing lanes or conduct additional evasive actions [14]. Because of limited smart sensors, some intelligent cars are unable to foresee the lane change moves of nearby vehicles in advance.

Smart cars identify lane-changing operations only after the target vehicle approaches or even crosses the lane markers. The target vehicle is supposed to be operated by a person in this example, which implies the person determines whether or not to make the turn. The driver can influence the vehicle's behavior in one way or another, as depicted in Fig. 7.1. Furthermore, the driver's decision is influenced by the vehicle's condition and dynamics. The target vehicle's behavior is determined by both the driver's input signal and the driving conditions, with the driver being the most important element in long-term behavior. However, lane change intentions are influenced by a variety of other factors, including neighboring traffic, type of road, road rules, and the driver's goal. The following are challenges that affect driver behavior.

1. **Driving behavior:** The driver's behavior, particularly the face and neck motion, is frequently used to assess the driver's purpose. Distractions, workload, multitasking, and exhaustion, among other factors, can lead to unintended behavior, which is not possible to forecast ahead of time. Because it is not possible to forecast the driver's intent in the approaching target vehicle, the lane-changing detection method relies primarily on the vehicle's movement to identify instead of forecast the lane shift. However, the vast majority of lane change movements are intentional. To support cognitive control of driving, a three-level hierarchy has been suggested: Strategic, tactical, and operational or vehicle control. The planning phase encompasses basic travel options, which include setting travel aims, selecting paths, and analyzing the financial risks of different travel options. The tactile level entails negotiating typical driving circumstances such as bends and crossings, accepting gaps when passing or joining the traffic, and avoiding obstacles. The functional layer consists of instant vehicle input signals, which are generally automated movement sequences.
2. **Traffic condition:** During driving, the driver is constantly monitoring the surroundings and traffic conditions. When another, more preferred lane becomes available, the driver will intend

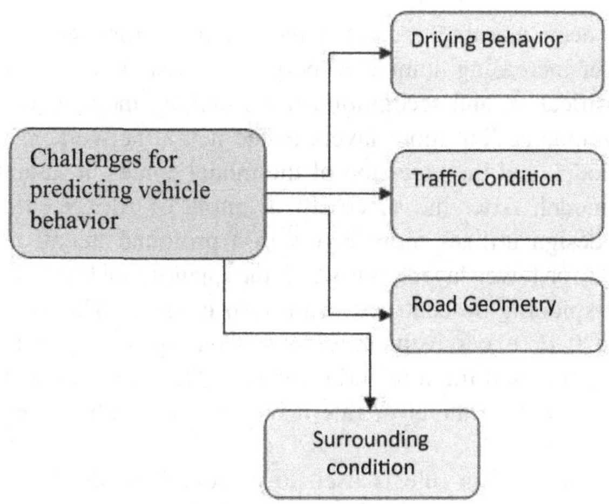

FIGURE 7.1

Important factors affecting vehicle behavior [14].

to go to that lane. The driver will then double-check his surroundings to ensure his security. Finally, the driver starts lane shifting, causing visible changes in the vehicle's state and travel data. The sensing capabilities of intelligent vehicles have a direct influence on the development and effectiveness of the lane-changing system. The majority of the environment perception is made up of nearby traffic, route details, and other environmental senses, as well as bordering congestion, which relates to people, cars, motorbikes, and so on. Road data contains lane design, marking styles, lane markings, and so on. Other sensory factors, such as traffic signals and weather conditions, have a substantial impact on driving habits.

3. **Road geometry:** Camera maps are commonly used to determine road geometry. Surrounding traffic can sometimes be used to approximate road shape. The same environmental data might originate from many sensing sources. Cameras and radars, for example, may detect nearby vehicles, and vehicle-to-infrastructure (V2I) communication and cameras can identify traffic lights.
4. **Surrounding conditions:** Paths and circular curvatures should be used in properly planned lane markers; it should be aligned, smooth, and unbroken with no sharp bends. The marks on the left and right are continuous, whereas the other markings are separated into segments. A unique and robust multiple lane identification approach was utilized based on route information, with five complementary limitations: Width, concurrent, distributed, and equal widths constraints [15]. Lack of path planning accuracy, modified and faded road lines when fresh markings are set, tire paw prints, low visibility, illuminations, ambient lights, strong shades, lane layout variances, and street directions remain challenging for lane marking detection. Moreover, lanes come in a variety of designs, such as smooth, dotted, painted, and unpainted markings.

7.2 ResNet architecture

Over the last decades, deep learning models have become progressively deeper (adding more layers) to address the ever-increasing number of perplexing assignments, which additionally helped in working out the classification and recognition tasks making them more powerful. In any case, when one can try to continue adding more layers to the neural network, it turns out to be particularly hard to train our model, and the precision of the model comes in saturation points resulting in the degradation of the model. After the successful winning of AlexNet in the 2012 competition, every ensuing winning design utilizes more layers in a profound neural network to decrease the error rate. AlexNet works for fewer layers, but when the quantity of layers is incremented, a typical issue called Vanishing/Exploding descent occurs in deep learning. This makes the inclination, i.e., the gradient, to become 0 or excessively enormous. Consequently, when the number of layers increases, it results in an increased training and error rate. This problem occurs during the backpropagation process. Hence, to reduce the error rate it is required to adjust weights and bias after calculating the loss/cost function.

During backpropagation, a chain rule is used to calculate the derivatives of every layer with respect to its previous layer function. At this point, by utilizing a number of more deep layers along with some hidden layers there is a possibility of scaling down the derivative size to some extent. When **n** number of layer derivatives are increased, the slope decreases dramatically, which leads down to the first layers. ResNet comes as a rescue from this situation and assists with settling this issue.

Consider an example in which one deep learning model is of 56-layers, and the other model has 20 layers. After processing, it was observed that models with 56-layers CNN generate more error, increasing the rate on both training and testing datasets than a 20-layer CNN design, and this was the aftereffect of over-fitting. In the wake of investigating more on the error rate, researchers had the option to arrive at the conclusion that it happened by vanishing gradient/exploding gradient.

To take care of the issue of the vanishing gradient/exploding gradient, some engineers presented the idea of a residual network (ResNet), which is one of the renowned deep learning models presented by Shaoqing Ren, Kaiming He, Jian Sun, and Xiangyu Zhang in their paper named *Deep Residual Learning for Image Recognition* [16] in 2016. The ResNet model is one of the most famous and best deep learning models. In this network the author uses a concept of skip connection in which it skips some of the training from the model (from a few of the layers) and directly provides results. The methodology behind this network performance is that rather than familiarizing the layers with the basic planning, the network is fitted with the residual mapping planning. Numerous residual squares are stacked together to shape a ResNet. Fig. 7.2 provides the proper working of a deep residual network with weights, activation functions, and residual blocks. The thought is to provide an interface of the input layer straightforwardly to the result layer by skipping a couple of layers. As per Fig. 7.2, x is the input to the layer, which is utilized to get the output by skipping a few layers associated with the model, and the result from such a connection will be $G(x)$. Then, at that point, the result will be $G(x) = f(x) + x$.

The benefit of adding this kind of skip association in such a case is that if any layer hurts the presentation of design, it will be skipped by regularization. During the operation there is a possibility that the dimension of the input may vary from that of the output, which happens with convolutional and pooling layers. Hence to handle such issues, two approaches are provided:

1. Padding of zeros with the skip connection to increase the dimensions in a model.
2. Or, the addition of one 1x1 convolutional layer to the input to match the dimensions and to calculate the output.

$$G(x) = f(x) + w1.x \tag{7.1}$$

The additional parameter is added in the second approach, which is actually not required while using the first approach.

Fig. 7.3 shows ResNet comprises one convolution and max-pooling layer followed by a number of convolution layers with the same behavior. First, the input matrix ($\times$ images) is provided to the network. Next, a convolution layer of 7×7 channels with a decent feature map aspect (f) [64, 128, 256, 512, 1024, 2048] is applied individually that bypasses the information of two convolutions. Moreover, the width (w) and height (h) aspects stay consistent during the whole layer.

The decrease in layers is accomplished by an expansion of the step from 1 to 2, at the primary convolution of each layer, rather than by a pooling layer, which is utilized to consider down samplers. ResNet architecture utilizes the CNN blocks multiple times, so a class for CNN blocks is created, which takes input channels and result channels. The subsequent stage is batch normalization, which is an element-wise activity, and thus, it does not change the size of our volume. At long last, the (3×3) Max Pooling activity with a stride 2 was used. The pooling layer in CNN reduces the

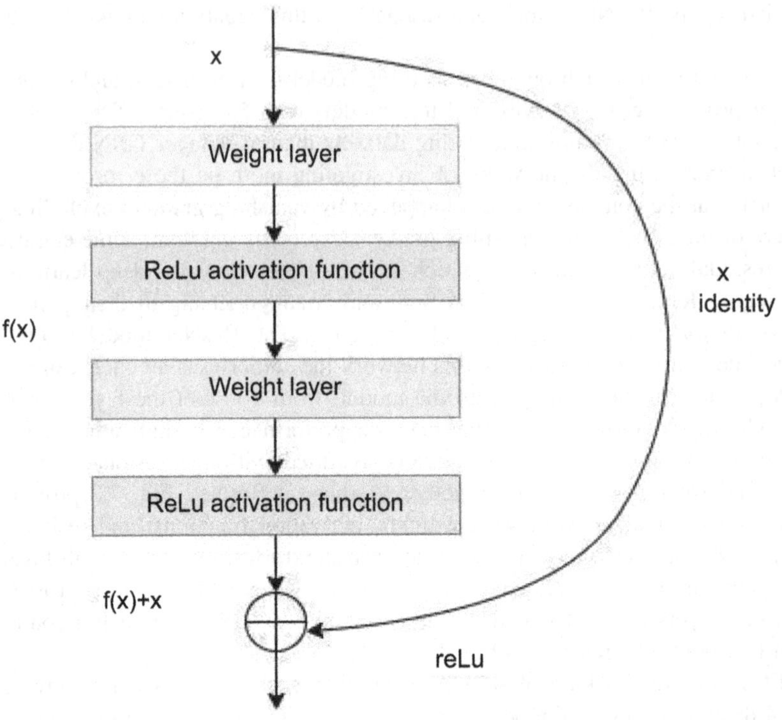

FIGURE 7.2

Functioning of a residual network.

dimensionality, making computation easier and providing faster training. With the help, it deduces the first pad info volume, so the last volume has the ideal aspects.

Calculating shape of feature map:

$$\text{Output height of feature map} = \frac{\text{Input height} - \text{filter height}}{\text{Row stride}} + 1 \tag{7.2}$$

$$\text{Output height of feature map} = \frac{\text{Input width} - \text{filter width}}{\text{Column stride}} + 1 \tag{7.3}$$

A fully connected neural network comprises a progression of fully connected layers. A completely associated layer is a function from $F\ x$ to $F\ y$. Each result aspect relies upon each info aspect. The fully connected neural network will take input of multiple dimensions and convert it into a single dimension, known as the flattening process. In a fully connected layer, for x inputs and y outputs, the number of weights is x^*y. Additionally, you have a bias for each output node, so a total of $(x + 1)^*y$ parameters. Fig. 7.4 provides detailed information about the working parameters of each and every layer.

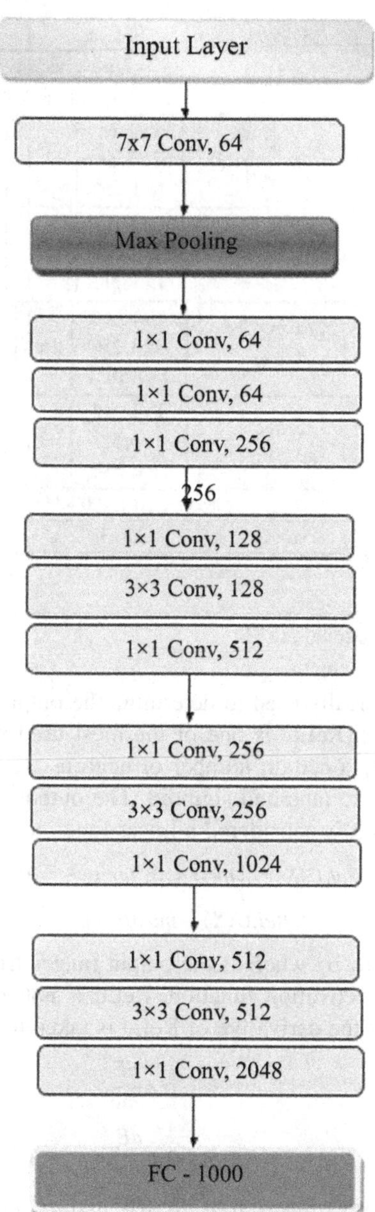

FIGURE 7.3

Architecture of a residual network ResNet.

layer name	output size	18-layer	34-layer	50-layer	101-layer	152-layer
conv1	112×112	7×7, 64, stride 2				
conv2_x	56×56	3×3 max pool, stride 2				
		$\begin{bmatrix} 3\times3, 64 \\ 3\times3, 64 \end{bmatrix}\times2$	$\begin{bmatrix} 3\times3, 64 \\ 3\times3, 64 \end{bmatrix}\times3$	$\begin{bmatrix} 1\times1, 64 \\ 3\times3, 64 \\ 1\times1, 256 \end{bmatrix}\times3$	$\begin{bmatrix} 1\times1, 64 \\ 3\times3, 64 \\ 1\times1, 256 \end{bmatrix}\times3$	$\begin{bmatrix} 1\times1, 64 \\ 3\times3, 64 \\ 1\times1, 256 \end{bmatrix}\times3$
conv3_x	28×28	$\begin{bmatrix} 3\times3, 128 \\ 3\times3, 128 \end{bmatrix}\times2$	$\begin{bmatrix} 3\times3, 128 \\ 3\times3, 128 \end{bmatrix}\times4$	$\begin{bmatrix} 1\times1, 128 \\ 3\times3, 128 \\ 1\times1, 512 \end{bmatrix}\times4$	$\begin{bmatrix} 1\times1, 128 \\ 3\times3, 128 \\ 1\times1, 512 \end{bmatrix}\times4$	$\begin{bmatrix} 1\times1, 128 \\ 3\times3, 128 \\ 1\times1, 512 \end{bmatrix}\times8$
conv4_x	14×14	$\begin{bmatrix} 3\times3, 256 \\ 3\times3, 256 \end{bmatrix}\times2$	$\begin{bmatrix} 3\times3, 256 \\ 3\times3, 256 \end{bmatrix}\times6$	$\begin{bmatrix} 1\times1, 256 \\ 3\times3, 256 \\ 1\times1, 1024 \end{bmatrix}\times6$	$\begin{bmatrix} 1\times1, 256 \\ 3\times3, 256 \\ 1\times1, 1024 \end{bmatrix}\times23$	$\begin{bmatrix} 1\times1, 256 \\ 3\times3, 256 \\ 1\times1, 1024 \end{bmatrix}\times36$
conv5_x	7×7	$\begin{bmatrix} 3\times3, 512 \\ 3\times3, 512 \end{bmatrix}\times2$	$\begin{bmatrix} 3\times3, 512 \\ 3\times3, 512 \end{bmatrix}\times3$	$\begin{bmatrix} 1\times1, 512 \\ 3\times3, 512 \\ 1\times1, 2048 \end{bmatrix}\times3$	$\begin{bmatrix} 1\times1, 512 \\ 3\times3, 512 \\ 1\times1, 2048 \end{bmatrix}\times3$	$\begin{bmatrix} 1\times1, 512 \\ 3\times3, 512 \\ 1\times1, 2048 \end{bmatrix}\times3$
	1×1	average pool, 1000-d fc, softmax				
FLOPs		1.8×10^9	3.6×10^9	3.8×10^9	7.6×10^9	11.3×10^9

FIGURE 7.4

Deep residual learning for image recognition [16].

The activation function is basically used to determine the output of a neural network model like 0 (false) and 1 (true). Among all, ReLU is one of the most used nonlinear activation functions in deep learning models. Since only a certain number of neurons are activated ReLU is computationally very efficient in comparison to tanh and sigmoid. The output of all the negative values is considered 0 and for positive values it is considered a linear function i.e., 1.

$$ReLU(X) = \{for\ x < 0;\ for\ x > = 0\} \tag{7.4}$$

$$ReLU(X) = max(0, X) \tag{7.5}$$

The input ranges from $-\infty$ to ∞ whereas the output ranges from 0 to ∞. But if one can compare ReLu to sigmoid and tanh activation function, ReLu is not differentiable for $x = 0$. To solve this for implementation purposes the derivative of ReLu is taken to be 0 at value 0.

$$w = w - \alpha * \frac{dE}{dw} \tag{7.6}$$

$$b = b - \alpha * \frac{dE}{db} \tag{7.7}$$

When we calculate the derivative function of dE/dW and dE/db, then

$$\begin{aligned} \frac{dE}{dw} &= \frac{dE}{d0} * \frac{d0}{dz2} * \frac{dz2}{dh1} * \frac{dh1}{dz1} * \frac{dz1}{dw} \\ \frac{dE}{db} &= \frac{dE}{d0} * \frac{d0}{dz2} * \frac{dz2}{dh1} * \frac{dh1}{dz1} * \frac{dz1}{db} \end{aligned} \tag{7.8}$$

The derivative of the activation function in this process was also calculated. When the derivative of the activation function becomes "0", then the bias value does not get updated; hence output becomes "0". Hence the weight will not update. Since only a certain number of neurons are activated, ReLu is computationally more efficient compared with other activation functions. Fig. 7.4 provides overall information about the number of filters, layers, activation function used, derivative function, and weights and bias used.

7.2.1 Dataset description

1. **TuSimple:** TuSimple is a commonly used route-finding database containing more than 6000 images, of which 3500 + are training images and 2700 + are test images of highly photographed highways in sunny and clear weather. However, the TuSimple data set contains video clips (where 20 frames were filmed per clip), and the final draft of each clip is noted. The results of the dynamic tests are shown in Table 7.1. A simple benchmark test code is analyzed with a database of two widely used open source benchmarks: SCNN [17] and LaneNet [18]. It is found that the combined measurement of slope boundaries and lines has been achieved with better acquisition accuracy of more than 3% externally, which confirms the need for slope consideration. Although the accuracy is worse than SCNN, it is important to note that both SCNN and LaneNet are trained with TuSimple data. Additionally, the designed method extracts trajectory parameters from the real world, and the findings are then mapped in contrast to the image view for comparison; but SCNN and LaneNet focus on channels in image viewing without disturbing the slope of the road or the variability of the camera parameters. Apart from this, the accuracy is comparably improved from 95.90% to 96.01% after consecutive accounting information. This seems to suggest that TuSimple's driving data was very common, and the standard auto-guessing was close to the real solution. In this study paper, temporary integration is done by considering only the output and the previous frame as the first prediction of the current framework. In real systems, multiframe integration can be large and enhanced, and temporary location filtering can also be used.
2. **CULane:** CUlLane is a major dynamic database of educational research on the acquisition of a traffic route. It was compiled by cameras installed in six different vehicles driven by different drivers in Beijing. More than 55 hours of video were collected, and 133,235 frames were released. The accuracy and analysis of the F1-score data are shown in Table 7.2. We divided the database into 88,880 for the training set, 9675 for the verification set, and 34,680 for the test set. In each frame, it defines traffic lines in cubic splines. In cases where the road markings have been blocked or unmarked, it still defines the routes in context, as indicated in Fig. 7.5A and B. It is also hoped that the algorithms can detect roadblocks. The routes on the other side of the barrier are therefore not defined. In this database the author focuses our attention on finding the four route signals, most of which are focused on actual applications. Some route features are not specified.

Table 7.2 CondLaneNet [19].

S. No.	Dataset	Task	Model Name	F1 Score	Accuracy
1	CULane	Lane detection	**CondLaneNet (ResNet-101)**	**79.48**	—
			CondLaneNet (ResNet-34)	**78.74**	—
			CondLaneNet (ResNet-18)	**78.14**	—
2	TuSimple	Lane detection	**CondLaneNet (ResNet-101)**	**97.24**	96.54
			CondLaneNet (ResNet-34)	**96.98**	95.37
			CondLaneNet (ResNet-18)	**97.01**	95.48

(A) TuSimple (B) CULane

FIGURE 7.5

Examples of video frames of (A) TuSimple [20] and (B) CULane [24]. Ground lane markers are shown through hue-colored lines.

Fig. 7.6 shows the localization of lane markers is successful at night, in the shadows, and when passing under the tunnel based on the model ERFNet uses end-to-end lane marker detection (E2E-LMD) (Figs. 7.7–7.9).

7.3 Conclusions

We have performed an analysis of the CUlane and TuSimple benchmark database based on three CondLaneNet versions. The results are shown in Table 7.1. We have considered the route acquisition model based on the actual conditional event classification strategy as a basis. In Table 7.1, the

FIGURE 7.6

E2E-LMD uses ERFNet as the backbone network in CUlane and TuSimple test images. All lines except the last one show CUlane's test images. Green dots are a good sample for visual purposes [21].

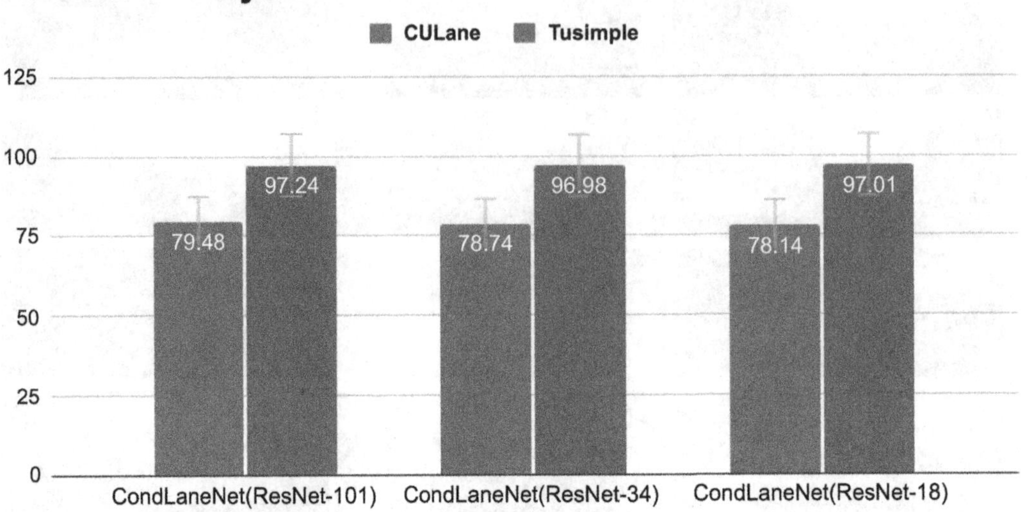

FIGURE 7.7

F1 score analysis for CULane and TuSimple on CondLaneNet (RESNet-101, RESNet-34, and RESNet-18).

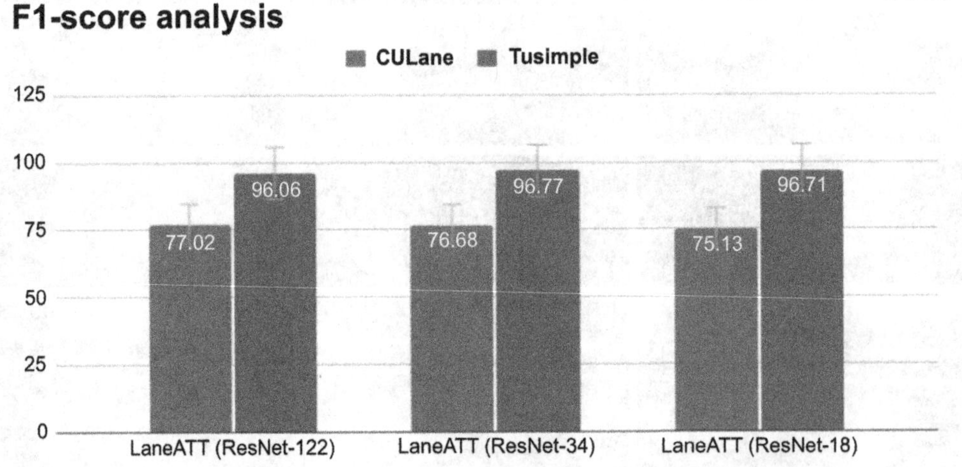

FIGURE 7.8

F1 score analysis for CULane and TuSimple on LaneATT (RESNet-101, RESNet-34, and RESNet-18).

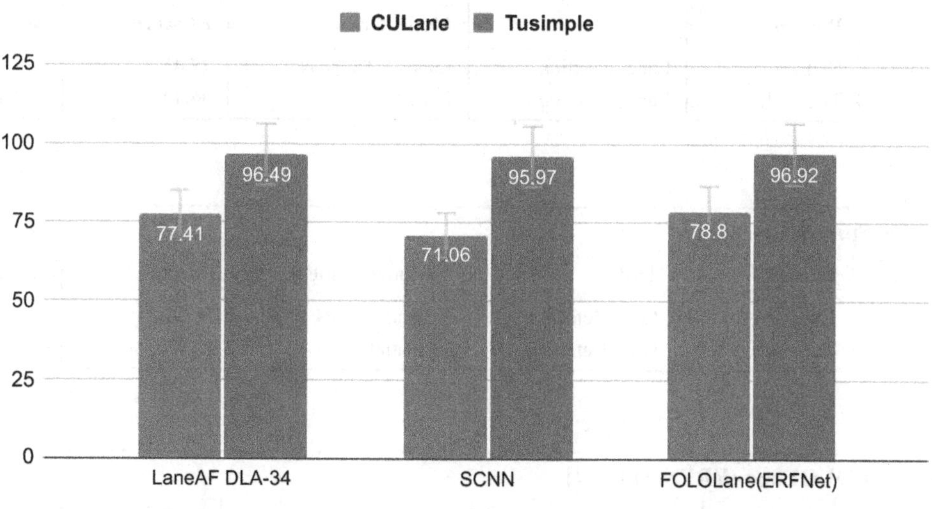

FIGURE 7.9

F1 score analysis for CULane and TuSimple on LaneAF, SCNN, and FOLOLane (ERFNet).

Table 7.3 LaneATT [B] [22].

S. No.	Dataset	Task	Model Name	F1 Score	Accuracy
1	CULane	Lane detection	**LaneATT (ResNet-122)**	**77.02**	—
			LaneATT (ResNet-34)	**76.68**	—
			LaneATT (ResNet-18)	**75.13**	—
2	TuSimple	Lane detection	**LaneATT (ResNet-122)**	**96.06**	96.10
			LaneATT (ResNet-34)	**96.77**	95.63
			LaneATT (ResNet-18)	**96.71**	95.57

first three lines show CUlane results for CondLaneNet based on versions ResNet-101, ResNet-34, ResNet-18. F1 score is recorded with accuracy and comparison. Similarly, the following three lines show the results for TuSimple, F1 score, and accuracy in the same hyperparameter set. Fig. 7.4 shows the F1-score analysis of the CondLaneNet model in CUlane and TuSimple. High accuracy and F1 score were noted in the CondLaneNet training model versions ResNet-101 in both cases. The similarity of the training model performed in the LaneATT versions ResNet-122, ResNet-34, ResNet-18 is shown in Table 7.2. It has been observed that F1 score and recorded accuracy were lower compared with the median accuracy for CondLaneNet model for route acquisition Fig. 7.5 shows the F1 score analysis of LaneATT models. Table 7.3, shows the F1-score visibility and accuracy in LaneAF DLA-34 and Culane LaneAF models and in the TuSimple data set. Similar

Table 7.4 LaneAF [23].

S. No.	Dataset	Task	Model Name	F1 score	Accuracy
1	CULane	Lane detection	LaneAF DLA-34	**77.41**	—
2	TuSimple	Lane detection	LaneAF	**96.49**	95.64%

Table 7.5 Spatial CNN [17].

S. No.	Dataset	Task	Model Name	F1 score	Accuracy
1	CULane	Lane Detection	Spatial CNN	**71.06**	—
2	TuSimple	Lane Detection	Spatial CNN	**95.97**	96.53%

Table 7.6 FOLOLane (ERFNet) [7,24].

S. No.	Dataset	Task	Model Name	F1 Score	Accuracy
1	CULane	Lane detection	FOLOLane (ERFNet)	**78.8**	—
2	TuSimple	Lane detection	FOLOLane (ERFNet)	**96.92**	96.92%

observations were recorded on the CNN spatial model as shown in Table 7.4. The finding of Table 7.5 shows that FOLOLane (ERFNet) works less closely than all other types of lane marker detection solutions. Fig. 7.6 shows the F1-score analysis of three different models, such as LaneAF, Spatial CNN, and FOLOLane (ERFNet) as mentioned in Table 7.6.

References

[1] D.C. Andrade, et al., A novel strategy for road lane detection and tracking based on a vehicle's forward monocular camera, IEEE Trans. Intell. Transport. Syst. 20 (4) (2019) 1497–1507.

[2] C. Wang, Q. Sun, Z. Li, H. Zhang, K. Ruan, Cognitive competence improvement for autonomous vehicles: a lane change identification model for distant preceding vehicles, IEEE Access. 7 (2019) 83229–83242.

[3] Z. Feng, M. Li, M. Stolz, M. Kunert, W. Wiesbeck, Lane detection with a high-resolution automotive radar by introducing a new type of road marking, IEEE Trans. Intell. Transport. Syst. 20 (7) (2019) 2430–2447.

[4] Q. Zou, H. Jiang, Q. Dai, Y. Yue, L. Chen, Q. Wang, Robust lane detection from continuous driving scenes using deep neural networks, IEEE Trans. Vehicular Technol. 69 (1) (2020) 41–54.

[5] S. Luo, X. Zhang, J. Hu, J. Xu, Multiple lane detection via combining complementary structural constraints, IEEE Trans. Intell. Transport. Syst. 22 (12) (2021) 7597–7606.

[6] P. Lu, C. Cui, S. Xu, H. Peng, F. Wang, SUPER: a novel lane detection system, IEEE Trans. Intell. Veh. 6 (3) (2021) 583–593.

[7] K. He, X. Zhang, S. Ren, J. Sun, Deep residual learning for image recognition, CVPR (2016).
[8] A. Gurghian, T. Koduri, S.V. Bailur, K.J. Carey, Deeplanes: end-to-end lane position estimation using deep neural networks, in: Proc. IEEE Comput. Soc. Conf. Comput. Vis. Pattern Recognit. Workshops, 2016, pp. 38–45.
[9] S.P. Narote, P.N. Bhujbal, A.S. Narote, D.M. Dhane, A review of recent advances in lane detection and departure warning system, Pattern Recognit. 73 (2018) 216–234.
[10] P. Viswanath, et al., A diverse low cost high performance platform for advanced driver assistance system (ADAS) applications, in: Proc. IEEE Conf. Comput. Vis. Pattern Recognit. Workshops (CVPRW), 2016, pp. 819–827.
[11] X. Zhang, W. Hu, N. Xie, H. Bao, S. Maybank, A robust tracking system for low frame rate video, Int. J. Comput. Vis. 115 (3) (2015) 279–304.
[12] R. Song, B. Li, Surrounding vehicles' lane change maneuver prediction and detection for intelligent vehicles: a comprehensive review, IEEE Trans. Intell. Transport. Syst. doi: 10.1109/TITS.2021.3076164.
[13] S. Luo, X. Zhang, J. Hu, and J. Xu, Multiple lane detection via combining complementary structural constraints, IEEE Trans. Intell. Transport. Syst. (2020).
[14] L. Liu, X. Chen, S. Zhu, P. Tan, CondLaneNet: a top-to-down lane detection framework based on conditional convolution, in: Proceedings of the IEEE/CVF International Conference on Computer Vision (ICCV), 2021, pp. 3773–3782.
[15] S. Yoo, H.S. Lee, H. Myeong, S. Yun, H. Park, J. Cho, et al., End-to-end lane marker detection via row-wise classification, in: CVF Conference on Computer Vision and Pattern Recognition Workshops (CVPRW), IEEE, 2020, doi: 10.1109/CVPRW50498.2020.00511.
[16] K. Zhang, M. Fu, Y. Yang, S. Shang, M. Wang, An efficient decision and planning method for high speed autonomous driving in dynamic environment, in: Proc. IEEE Intell. Vehicles Symp. (IV), 2017, pp. 806–811.
[17] H. Abualsaud, S. Liu, D. Lu, K. Situ, A. Rangesh, M.M. Trivedi, LaneAF: Robust multi-lane detection with affinity fields, IEEE Robot. Autom. Lett. 6 (4) (2021). Available from: https://doi.org/10.1109/LRA.2021.3098066.
[18] L. Tabelini, R. Berriel, T.M. Paixão, C. Badue, A.D. Souza, T. Oliveira-Santos, Keep your eyes on the lane: real-time attention-guided lane detection, in: IEEE/CVF Conference on Computer Vision and Pattern Recognition (CVPR) 2021. doi: 10.1109/CVPR46437.2021.00036.
[19] X. Pan, et al., Spatial as deep: spatial CNN for traffic scene understanding, in: Proc. AAAI, New Orleans, LA, United States, February 2–7, 2018, pp. 7276–7283.
[20] TuSimple, Tusimple lane detection benchmark, 2017. https://github.com/TuSimple/tusimple-benchmark.
[21] N. Davy, et al., Towards end-to-end lane detection: an instance segmentation approach, in: Proc. IEEE Intell. Veh. Symp., June 2018, pp. 286–291.
[22] P. Lu, C. Cui, S. Xu, H. Peng, F. Wang, SUPER: a novel lane detection system, IEEE Trans. Intell. Veh. 6 (3) (2021).
[23] Z. Qu, H. Jin, Y. Zhou, Z. Yang, W. Zhang, Focus on Local: Detecting Lane Marker from Bottom Up via Key Point, in: IEEE/CVF Conference on Computer Vision and Pattern Recognition (CVPR), 2021. doi: 10.1109/CVPR46437.2021.01390.
[24] Z. Tian, C. Shen, H. Chen. Conditional convolutions for instance segmentation, in: Proceedings of the European Conference on Computer Vision (ECCV), 2020.

CHAPTER

Intelligent autopilot fire extinguishing robot

M.N. Sumaiya[1], J. Vineeth[1], Prashanth Sali[1], G.R. Supreeth[1] and R. Supreeth[2]

[1]*ECE Department, Dayananda Sagar Academy of Technology & Management, Bengaluru, Karnataka, India* [2]*Anglia Ruskin University, Cambridge, United Kingdom*

8.1 Introduction

In the last two decades, the purpose of robots in industries in automating industrial processes has been extended. The processes include fabrication, finishing, transferring, and assembling parts; distribution of finished products; picking, sorting, packaging, and palletizing products in manufacturing and material handling units. Numerous studies have shown that robots can be beneficial in aerospace, automotive, computers, e-commerce, medicine, rehabilitation, rescue operation, and industry. Industrial robots are multifunctional manipulators designed for more specialized materials, divisions, gadgets, or devices through various programmatic movements to perform various tasks. Still, there is a demand for a system that can control, communicate, and integrate different robots regardless of their types and specifications per Fourth Industrial Revolution (4IR) standards.

Recently, machine learning has also heated up interest in robotics to increase the intelligence of robots and the productivity in industry to reduce cost. Researchers got attention in designing humanoid robots to minimize firefighters' injuries and deaths as well as increase productivity, safety, efficiency, and quality of the task given. The types of robots used in different categories are telerobots, telepresence robots, mobile robots, autonomous robots, and android robots. Compared with telerobots, telepresence robots provide feedback from video, sound, and other data, and so it is widely used in fields requiring monitoring, such as in child nursery and education, and in improving older adults' social and daily activities. Mobile robots are designed. Autonomous robots, which can perform tasks independently and receive power from the environment, are created to mimic humans. Numerous types of vehicles for firefighting at home and extinguishing forest fires exist; still, there is a need for robots to be able to work on their own or be controlled remotely to extinguish the fire. In this paper, a cost-effective, compact-sized, unmanned support vehicle firefighting robot is proposed, which is designed and constructed to search and extinguish the fire in narrow spaces during critical situations.

8.1.1 Need for automatic fire extinguishing robot

During critical fire accidents in places like homes, industries, nuclear power plants, petroleum refineries, and gas tanks, people and properties exposed to higher risks are firefighters, humans in

Intelligent Edge Computing for Cyber Physical Applications. DOI: https://doi.org/10.1016/B978-0-323-99412-5.00002-2

that building, and the occupied space. According to statistics from the National Interagency Fire Center, there were 59 k fire accidents and 10.1 million acres burned down in 2020 around the world, double the space burned compared with the previous year's data. During fire disasters, many humans lose their lives. As per the ADSI-2019 report, there were 11,037 fire accidents reported across India in 2019. Since these disasters are increasing day by day, a cost-effective fire extinguishing robot has become obligatory in all such places. It is a challenging task to design fire extinguishing robots for hazardous situations, considering the size, material, and technology involved. The robot has to be equipped with sensors to detect the fire and remove obstacles from the path while searching for the fire. After locating the source of the fire, the robot should be able to extinguish the fire. Also, it should have the facility to allow the remote user to monitor, operate, and control the robot through a wireless communication system. Additionally, it has to face the hindrance of entering narrow and restricted places to extinguish the fire. To rescue buildings and objects from getting destroyed, compact-sized robots are required. Technological innovations can be used efficiently to facilitate highly responsive firefighting tasks. This work involves the development of a firefighting robot to avoid unnecessary dangers.

8.1.2 Existing methods

Swetha Sampath [1] proposed a robot that turns on automatically when it detects a fire within a distance of 5–10 cm using a thermocouple. The robot was experimented at a temperature of 300°C, which tried to extinguish the fire by moving in the direction of fire intensity. This technology does not support current technologies of monitoring and controlling from a remote place. Teh Nam Khoon et al. [2] developed an autonomous firefighting mobile platform and showed it to be a feasible project. Based on the findings, by integrating all the hardware, such as flame sensors, motor driver circuitry, and light-dependent resistor (LDR) sensors, the expected patrolling and fire extinguishing tasks can be carried out and executed with a minimum level of error. Additional features can be integrated into the system, namely the wireless communication module so that it can communicate between the operator and the victims within the fire site. Mangayarkarasi [3] proposes a system that uses an RF remote for remote operation along with RF receiver-based microcontroller circuit for operating the robot and water pump. This project can be developed by interfacing it with a wireless camera so that the person can view the controlling operation of the robot from a remote display. Alif et al. [4], QRob is designed to be more compact in size than other conventional firefighting robots to ease small location entry for deeper reach in extinguishing fire in a narrow space. QRob is also equipped with two sensors, an ultrasonic sensor to prevent it from hitting any obstacle and surrounding objects and a flame sensor for fire detection. The sensors enable QRob to identify fire locations automatically and to extinguish the fire remotely from a certain distance. QRob is programmed to find the fire location and stop at a maximum distance of 40 cm from the fire. A human operator can monitor the robot by using a camera connected to a smartphone or remote device. However, QRob does not have a backup system for more water supply. Garu et al. [5] designed a robot that can extinguish the fire before it rages out of control with minimum human intervention. The main objective was to build a framework to identify and quench the fire before the fire expands and creates any hazardous situation.

8.1.3 Scope and objective

Fire causes huge damage and loss to human life and property. It is sometimes impossible for firefighters to access the place of fire because of explosive materials, smoke, and high temperature. It is observed that robots can be more reliable for this job where human lives are at risk. Robots can be used to reach the environment in a short time, which is beyond human access. In such environments, autopilot fire extinguishing robots (AFER) can be useful in extinguishing a fire. Fire extinguisher robots are designed to be used in such extreme conditions. It can be operated and controlled by remote users with the ability to extinguish a fire after locating the source of the fire. Fire extinguisher robots are provided with a monitoring system and work on wireless communication technology. The fire detection system is designed using the sensors mounted on the fire extinguisher robot. Environmental fire is detected by the flame sensor, and sensed signals are transmitted to the microcontroller to trigger the pump, which sprinkles water to extinguish the fire. This robot is controlled using a mobile phone through a Bluetooth module. The information on flame, temperature, smoke, and presence of obstacles in the place of accident is detected by the sensors and processed by the Arduino Uno board; the processed information is then activated by the robot.

8.1.4 Literature survey

The evolution of robots from its start till the present day has been surveyed [6–10]. Craig's *Introduction to Robotics* [6] explains the history of industrial automation characterized by periods of rapid changes in popular methods. As a cause or, perhaps, as an effect, such a series of changes in automation techniques seem closely tied to world economics. The use of these unique devices, along with different design techniques such as computer-aided design (CAD) systems and computer-aided manufacturing (CAM) systems, characterizes the latest trends in the automation of the manufacturing process, saving time and human errors. The paper on the *Evolution of Robotics Research* [7] provides an overview of the evolution of research topics in robotics from classical motion control for industrial robots to modern intelligent control techniques. The introduction of industrial robots led to robotics research that has evolved toward the development of robotic systems to assist humans in dangerous, risky, or hazardous chores. With more and more complexities with these events, flexibility has been demanded in industrial robots, and robotics research has been led toward an adaptive and intelligent system that helps replace humans. In *Real-Time Obstacle Avoidance for Manipulators and Mobile Robots* [8], Khatib points out it is necessary to take care of obstacles that lie in the way or track of a moving object or vehicle. Khatib's paper presents an approach for manipulators and mobile robots based on the artificial potential field concept to achieve obstacle avoidance. They have described the formulation and implementation of this approach based on the artificial potential field concept. Since collision avoidance is generally treated as high-level planning, it has been demonstrated to be an effective component of low-level real-time control in this approach. Thus, they have briefly presented their operational space formulation of manipulator control that provides the basis for this obstacle avoidance approach and have described a two-level architecture designed to increase the real-time performance of the control system. The papers on *Robot Introduction in Human Work and Environment* and *Design and Implementation of Common Platform for Small Humanoid Robots* [9,10] suggest understanding common safety problems with robot installation in human working environments. We have seen robots used in applications to offer economic advantages, increased productivity, and higher consistency. With this kind of automation, manipulation of robots in the human environment would be risky as it may lead to technical failure. Hence they suggest sharing components and introducing standard frameworks to provide safety and progress in general. Also, these papers suggest having a common platform for all humanoid robots being used for research to avoid inefficient methodology.

The evolution and history of cyber-physical systems have been explained in different works [11–15]. *Cyber-Physical Systems—Concept, Challenges and Research Areas* [11] describes the nature and importance of cyber physical systems (CPSs), which play a major role in the design and development of future engineering systems where controlling of mechanical systems is done using computing paradigms, which also includes areas such as generic architecture, design principles, modeling, dependability, and implementation. These systems have developed to represent the integration of information and knowledge into physical objects. The main aim of these systems is to develop CPSs that are event-driven multiagent models that can combine the physical and cyber components and facilitate the correct study of their interdependencies. In *Cyber-Physical Systems: A Confluence of Cutting-Edge Technological Streams* and *Cyber-Physical Systems: A New Frontier* [12,13], the authors have worked in close coordination with the physical components and computational entities as a closed-loop control system through networked communications, which emerges as a great potential application. The above said work describes the characteristics, state-of-the-art research, challenges, and opportunities for solving complex application-level problems in CPS. The authors illustrate the concept of CPS with an example problem of tracking a bio-chemical weapon being carried by a terrorist. Wireless communication might encounter inaccurate measurements; CPS makes use of inaccurate or incomplete data from sensor networks to make intelligent control decisions to operate the actuators for effective control of physical processes. Sha et al. [13] discuss optimization, data transmission, and network security in wireless communication systems to show that the prospect of CPSs is prominent. In the *Security Aspects of Cyber Physical Systems* and *Cyber-Physical System Security and Impact Analysis* [14,15], the authors have discussed security threats imposed on CPSs by attackers to spoil them.

Security threats are of two types—control security and information security. Based on this division, various attack methods and their possible solutions have been discussed here. Illustrations are done by considering a smart home, consisting of several sensors, actuators, and cyber controllers, as all these devices are connected to a cyber system via the Internet or local net. As these devices share a lot of data over a server on the net, security concerns are raised regarding personal life involving personal information.

The paper on *The Role of Edge Computing in Internet of Things* [16] signifies the use of edge servers or edge computing to process data on the edge of the network as many IoT devices rely on computing paradigms for improving productivity and revenues. It also facilitates processing delay-sensitive and bandwidth-hungry applications near the data source. The paper by Liu et al. [17] is a work on blockchain-based video streaming systems that help in building decentralized networks with flexible monetization mechanisms for video streaming services. Video transcoding, which is computationally intensive and time-consuming, is still a major challenge on these blockchain-based platforms. Hence this paper proposes a novel blockchain-based framework with an adaptive block size for video streaming with mobile edge computing. Surveys [18,19] suggest that several new computing applications, such as virtual reality and smart environments, have become possible due to the availability of cloud resources and services. The cloud computing paradigm is unable to meet the requirements of low latency, location awareness, and mobility support. Therefore mobile edge computing (MEC) was introduced to bring cloud services and resources closer to the user by leveraging available resources in the edge networks so that the applications can process the data with greater accuracy and efficiency. The article on *Edge Computing in IoT-Based Manufacturing* [20] proposes an architecture of edge computing for IoT-based manufacturing. It also analyzes the role of edge computing from four aspects, including edge equipment, network communication, information fusion, and cooperative mechanism with cloud computing. This article aims to provide a technical reference for the deployment of edge computing in the smart factory as well as a system architecture for implementing edge computing in IIoT applications. Edge computing in IIoT meets real-time requirements of lightweight intelligent manufacturing, increasing the agility and security of the network. Edge computing provides obvious advantages in terms of business agility and bandwidth optimization compared with traditional approaches.

Kristi et al. [21] developed intelligent firefighting tank robots, consisting of compass sensors, flame detectors, thermal array sensors, white detectors (IR and phototransistors), sound activation circuits, and micro switch sensors to sense and extinguish the fire. The main objective of the robots is to search a certain area, and find and extinguish the flame for different flame positions. Singh et al. [22] developed an autonomous industrial firefighting mobile robot. They described the construction and design of mobile firefighting robots, containing isolated direct current (DC) motors, water pumps, and water containers to perform the task. The robots perform analog to digital conversion of the data provided by five infrared sensors, out of which two sensors control the motion of the robots, and three of them are for flame detection. The objective is to make the robots sense the flames of the fire, making use of the infrared sensor as an input sensor that helps in sensing the infrared rays coming out of the fire and the ability to extinguish it in that particular area of occurrence.

A wireless firefighting robot developed by Swathi Deshmukh [23] has the ability to detect fire and extinguish it using light-dependent resistors. The light-dependent resistors are used for the detection of fire, and the resistors are highly sensitive devices capable of detecting even a very small fire. This robot is developed with the concern of security at home, buildings, factories, and laboratories and is an intelligent multisensory-based security system that also contains a firefighting system.

A cell phone-controlled robot developed by Lakshay Arora [24] consists of fire detection sensors and a mobile phone to control a robot by making a call to the mobile phone connected to the robot. Whenever the call gets activated, the tone corresponding to the mobile phone attached performs actions accordingly. This particular system uses a dual-tone multiple-frequency (DTMF) technology, which is used to position the motor shaft at the required point with different sensors, each performing its own task. Arpit Sharma [25] developed an android-phone-controlled robot using Bluetooth. The aim is to capture gestures using an accelerometer and Bluetooth module to control the kinetic motion of the robot. Signals from the sources are controlled by a microcontroller, which helps in performing actions for the defined inputs.

Saravanan [26] has designed and developed an integrated semiautonomous firefighting mobile robot, which is controlled autonomously by a navigation system comprising infrared and ultrasonic sensors as well as a wireless camera to capture the videos and transmit them to the base station.

With the help of graphical user interface (GUI) support the robot can be controlled from the base station. LDR and temperature sensors are used to detect the fire. An intelligent fire extinguisher system is developed by Sonsale et al. [27], and their paper proposes an adaptive fusion algorithm for fire detection. This robot makes use of the smoke sensor, flame sensor, and temperature sensor for fire detection. It also contains an intelligent multisensory-based security system that helps in the detection of abnormal and hazardous situations occurring at places and notify them. Being an intelligent system, it cuts off the electricity of the area where fire has been caught and starts sprinkling water only in that area.

A remote-controlled firefighting robot developed by Phyo Wai Aung [28] describes the functions of the robot that contains a transmitter and a receiver in which two sets of RF (radio-frequency) modules are used. One of the modules is used to transmit the data to the motor driver, and the other is used to know the condition of the fire. The motors connected to the microcontroller are driven by L298 and ULN2003 drivers in this system. Robot movements are controlled by using a wireless camera mounted on the robot. If the temperature of the fire sight is above 40°C, the alarm starts ringing so that the operator can control the firefighting robot and avoid the damage of heat. In *Fire Fighting Robot Controlled Using Android Application* [29] the robot is remotely controlled using an android application. It makes use of two sensors to detect the fire – the smoke sensor (light intensity) and temperature sensor. It also makes use of commands, such as moving forward, left, and right, sent to the robot by using an android device. At the receiving end, two motors are interfaced to the microcontroller, one for the movement of the robot and the other to position the arm of the robot.

Fire Fighting Robot: An Approach [30] suggests the development of fully autonomous robots and implements the concepts of environmental sensing and awareness and proportional motor control. It uses an microcontroller manufacturer product number (SMCL) microcontroller to process information from its various sensors and hardware elements. It also makes use of ultraviolet, infrared, and visible light to detect various components of its environment.

Although a lot of research has been done in the field, the requirement for a fire extinguisher robot with technological advancements still exists. In this chapter, an intelligent AFER is proposed and developed. It is equipped with four different sensors, which help the robot to detect fire at longer ranges and is 60% more accurate than other robots. A flame sensor is used to sense environmental fire and feed the signals to the microcontroller to trigger the pump to sprinkle water to extinguish the fire. The robot is designed to have wheels with large belts so that it can move easily on any terrain. It has a water backup system, which other robots do not have. The robot is controlled manually or by using a mobile phone through Bluetooth. It has a 360-degree rotating robotic arm, which other robots do not have. The robot processes information from its various key hardware elements, such as the flame sensor, temperature sensor, smoke sensor, and ultrasonic sensor, via Arduino Uno board.

8.2 Proposed system: materials and methodology

The proposed system is classified into mechanical schematics, hardware description, and programming design. All these parts were assembled, and experiments were performed to determine the optimal distance of the proposed robot to extinguish the fire.

8.2.1 Block diagram

The rough sketch of the proposed robot is shown in Fig. 8.1. Based on the requirements, hardware components and technologies were combined, as shown in Fig. 8.2. Google SketchUp software and AutoCad were used to construct 3D and 2D schematic diagrams. The proposed robot has two wheels on the rear side and two wheels on the front side to obtain the chosen movement and speed to frame the main structure of the robot. For the wheels to stabilize the robot and make rotations, 360-degree rotation is used. The body of the robot is made of a heat-resistive acrylic plate to protect the electronic circuit. The acrylic sheet is resistant to the heat of up to 200°C. The body of the acrylic chassis contains holes that make it easier for the mounting of various types of sensors and other mechanical components. An ultrasonic sensor is installed in the front of the robot to avoid hitting obstacles. The flame sensor is used to detect fire. A mini camera installed on the front side of the robot is linked to a smartphone to monitor the location.

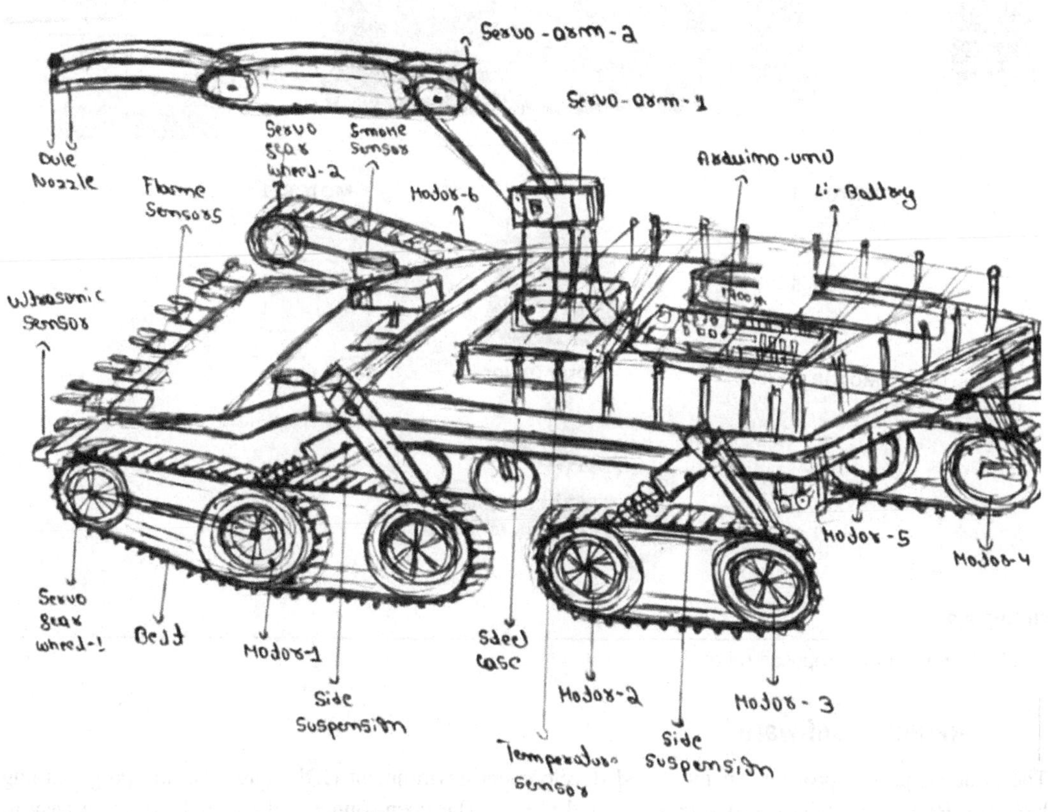

FIGURE 8.1

Rough sketch of the robot.

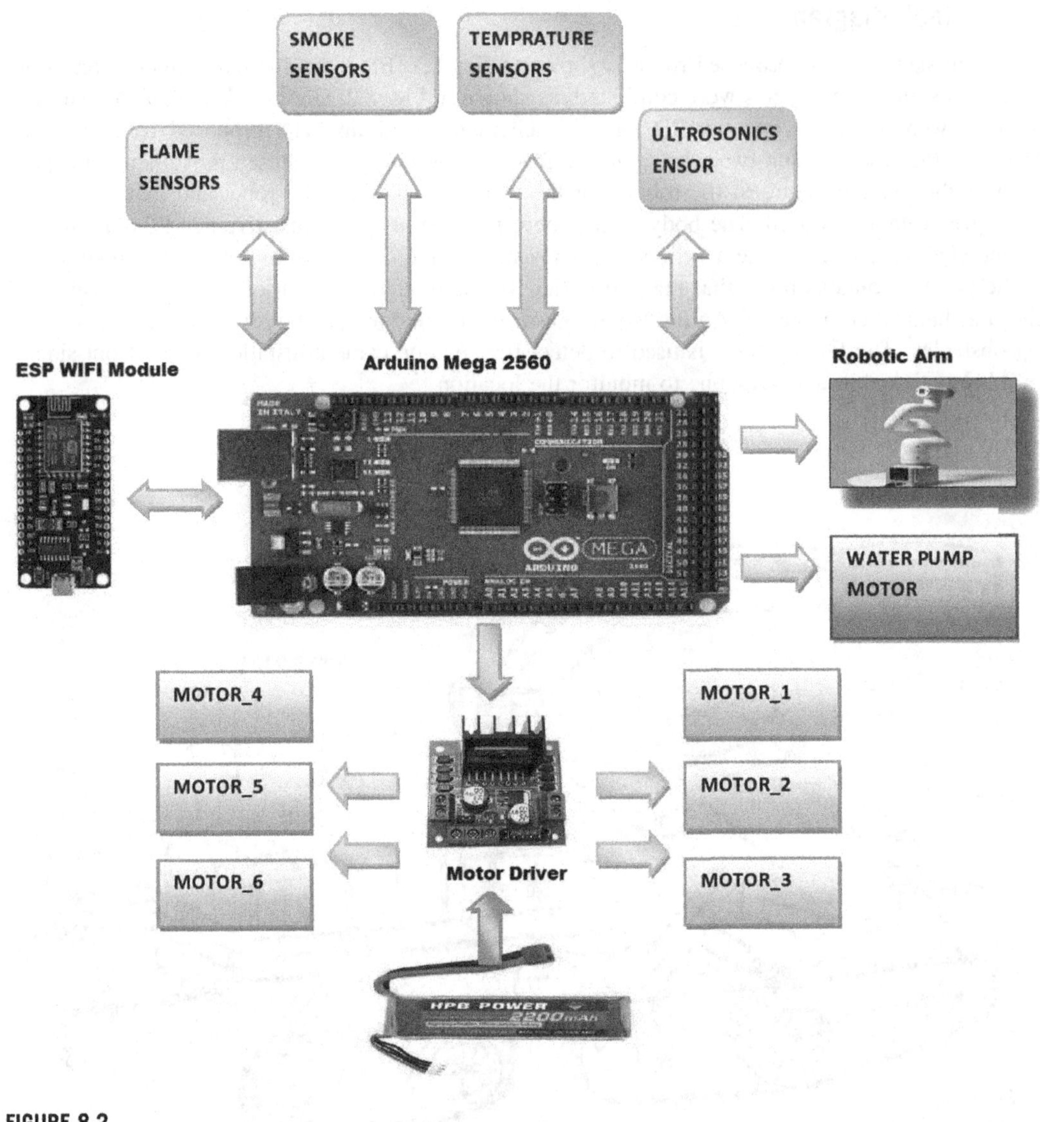

FIGURE 8.2

Block diagram of the proposed robot.

8.2.2 Required software

The Arduino project provides an integrated development environment (IDE) based on the programming language *Processing*, which also supports C and C++. The open-source Arduino IDE makes it easy to write the code and upload it to the board. Embedded C is a generic term for a programming language written in C, associated with a particular hardware architecture. Embedded C is an extension to the C language with additional header files, which change from controller to controller. Embedded C programming requires nonstandard extensions to the C language to support features such as fixed-point arithmetic, multiple distinct memory banks, and input/output operations. Arduino board designs use a variety of boards and controllers. The boards are equipped with sets of digital and analog input/output (I/O) pins that may be interfaced to various expansion boards or breadboards (for prototyping) and other circuits.

8.2.3 Arduino software

It is an easy-to-use hardware and software, which is primarily an open-source electronics platform used for many purposes. It can use different sets of inputs like a finger on a button, light on any sensor, etc. The Arduino project provides an IDE based on Processing language. The Arduino IDE supports C and C++ using special rules of code structuring. A text editor for writing codes, messages, text consoles, toolbars with different buttons for functions and a series of menus all form part of the IDE. Arduino uses the concept of a sketchbook – a standard place to store programs; the first-time use of Arduino software will automatically create a directory for the sketchbook. Each processor is, in turn, somehow connected to embedded software, which decides and is responsible for the functioning of the embedded system. Microcontrollers are most significantly programmed using embedded C language. The two most important features of embedded programming are code speed and code size. Processing power and timing constraints determine the code speed and program memory available, and which programming language is used determines the code size in the embedded system. The aim of embedded system programming is maximum features in minimum space and time. It has several advantages like low power consumption, rugged operating ranges, and low per unit cost. These systems are based on a microcontroller. Embedded C programming requires nonstandard extensions to the C language to support features such as fixed-point arithmetic, multiple distinct memory banks, and input/output operations. Embedded C is a generic term for a programming language written in C, associated with a particular hardware architecture. Embedded C is an extension to the C language with additional header files.

8.3 Implementation

This module can work both as an access point and as a station; hence it can easily transmit data to the Internet making IoT as easy as possible. It also fetches data using application programming interfaces (APIs), helping devices or systems access any information on the Internet. The programming of this module is done using Arduino IDE, which makes its features more exciting. Some applications include smart home applications, wireless data logging, portable electronics, etc. A fire extinguisher robot with a robotic arm has three sections: Obstacle detection; temperature, flame, and gas sensor detection; robot direction control. The flowchart of the proposed work is given in Fig. 8.3.

8.3.1 Obstacle detection

The block diagram, flow chart, and schematic diagram of obstacle detection are shown in Fig. 8.4, in which an ultrasonic sensor senses the obstacle and sends a signal to Arduino, which generates the command for the robotic arm to clear the path.

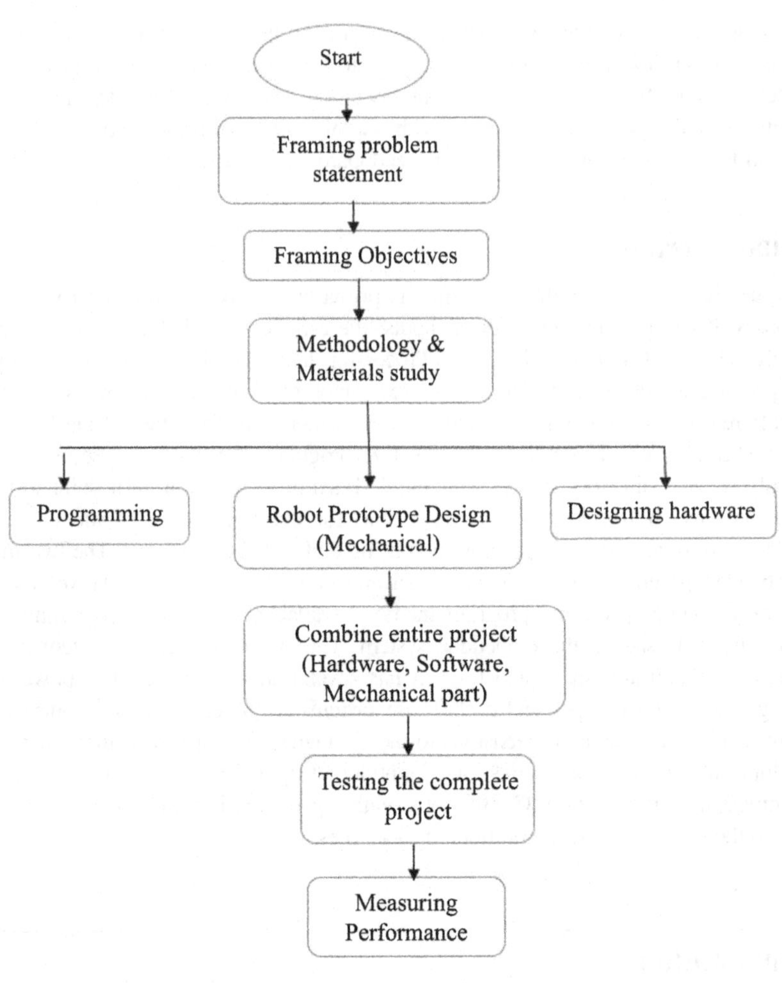

FIGURE 8.3

Flowchart of the proposed work.

8.3.2 Temperature, flame, and gas sensor detection

The block diagram and flow chart of flame, gas, and temperature sensors are shown in Fig. 8.5, in which the temperature sensor detects the temperature of the environment, and if the temperature crosses its limits, it activates the flame sensor to sense the fire. The flame sensor sends the signal to Arduino, after which Arduino generates the command for the water sprinkler to put the fire off from the desired location.

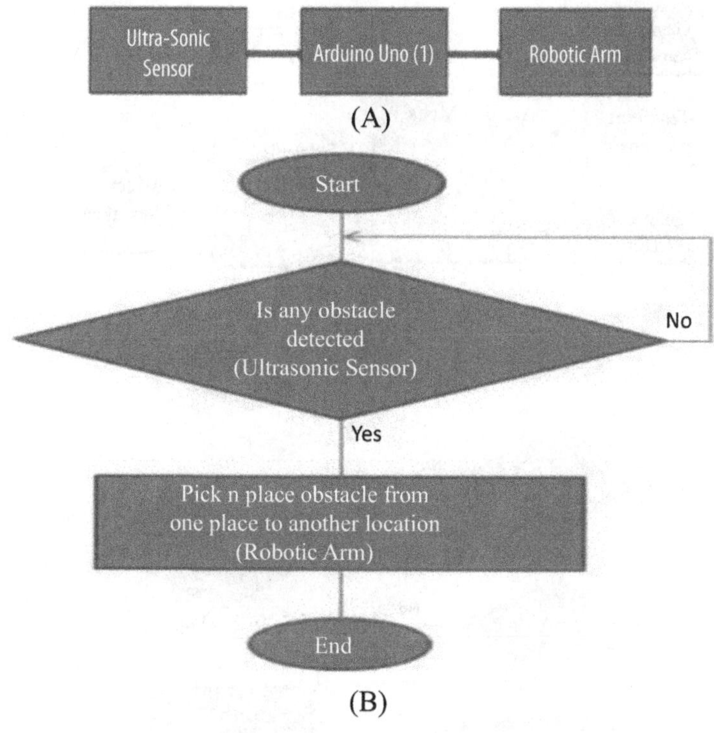

FIGURE 8.4

Flowchart of (A) obstacle detection section (B) functional flow.

8.3.3 Robot direction control

The block diagrams of video feedbacking and robot movement are shown in Fig. 8.6, in which an android cell phone is connected to the laptop via a mobile internet protocol (IP) camera application connected with its IP address of the desired location. Furthermore, the RF remote control controls the robot's direction – the robotic car can be moved in any direction.

8.3.4 Components involved

The hardware part is one of the imperative components in the development of the proposed robot, which includes several types of sensors, microcontrollers, DC motors with wheels, transmitters and remote controls, and water pumps. Fig. 8.2 shows the block diagram of the proposed robot. The diagram shows that various information like flame, smoke, temperature, and obstacle objects in and around the fire accident zone is sensed by the flame sensor, smoke sensor, temperature sensor, and

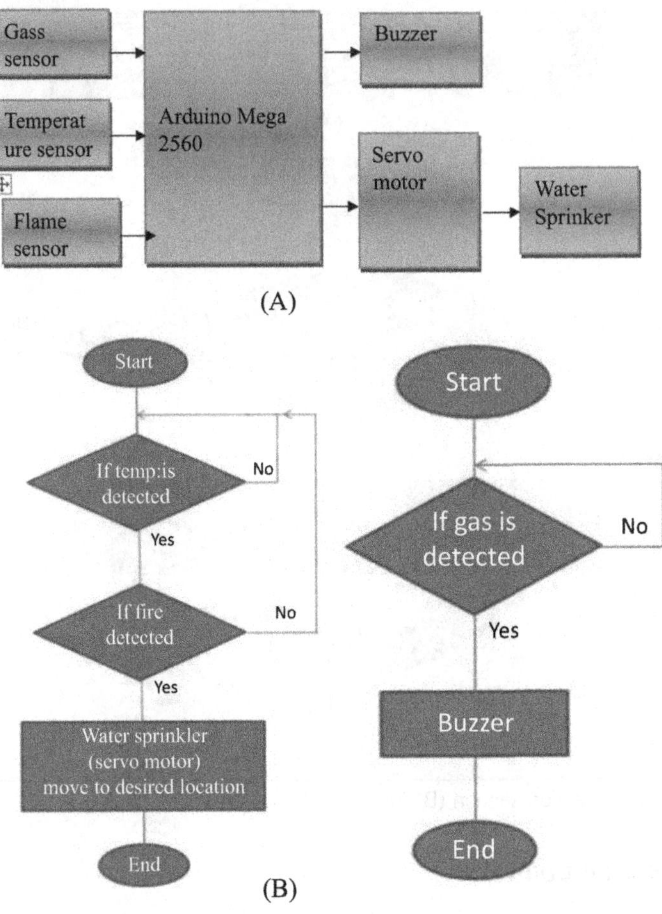

FIGURE 8.5

Flow chart of (A) flame, gas, and temperature sensors (B) functional flow.

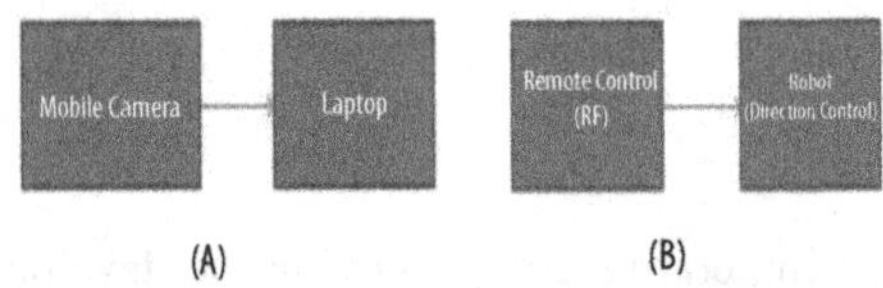

FIGURE 8.6

Flow chart of (A) video feedbacking (B) robot movement.

ultrasonic sensor, respectively. The sensed signals are sent as inputs to Arduino Uno, which processes the received signals to produce controlling signals for the appropriate devices. All hardware components are connected to Arduino Uno. To activate the moving of the gear motor, the motor driver (L298N) is connected to the central unit, while the transmitter remote control will provide the output of the system. The remote operator can control the flow of water and the fire extinguisher pump. On the other hand, the robot's movements can be monitored by the remote operator using a camera connected to a smartphone. The required components for implementing this project are listed in Fig. 8.7.

8.3.4.1 Flame sensor

The flame sensor, which identifies the origin of the fire first, plays an imperative role in most fire-fighting robots and acts as an eye for a robot. It has two signal pins—digital output (DO) and analog output (AO). The DO pin gives two information – there is flame or no flame, while the AO pin detects the exact wavelength of different lights. When the wavelength of light is at 760–1100 nm, the robot can identify fire effortlessly. The detection angle is roughly 60 degrees, and the distance is 20 cm (4.8 V) to 100 cm (1 V).

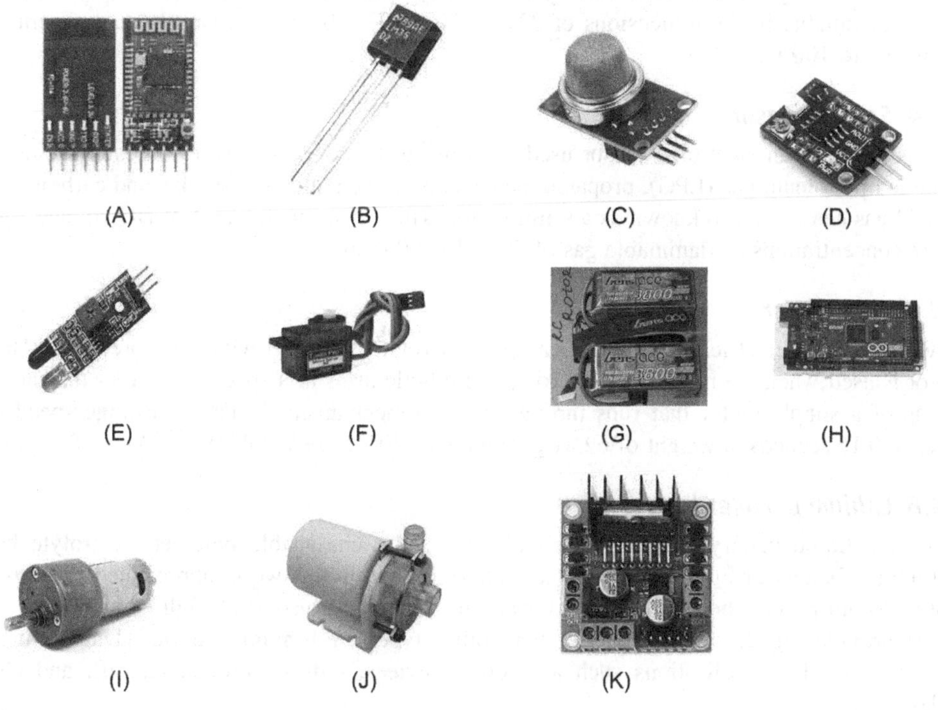

FIGURE 8.7

Required components (A) Bluetooth module (HC-05), (B) temperature sensor (C) smoke sensor (D) ultrasonic sensor (E) infrared sensor (F) servomotor (G) lithium polymer battery (H) ATmega2560 microcontroller (Arduino Uno) (I) geared motor (J) water pump (K) motor driver.

8.3.4.2 Ultrasonic sensor

The main hindrance to the automatic detection of fire is obstacles on the path while searching the fire. The obstacle has to be removed from the path to improve the performance of AFER. The sensor must be simple, compact, low cost to produce accurate sensing data values and be functional on a larger scale. Moreover, it should be smart to sense obstacles with enough boundaries to enable robots to respond and move in the appropriate direction. The existing sensors that suit all these requirements are ultrasonic sensors. The HCSR04 ultrasonic sensor is used in this study to detect obstacles within the range of 2–10 cm. However, it rotates at an angle of 360 degrees after a particular range is detected. This sensor transmits electromagnetic waves and receives reflected waves from the object to measure the location of the object. It has four output pins such as a reference voltage (Voltage Collector Collector (Regulated DC supply voltage (VCC))) (which operates around 5 V), ground pin (GND), digital output (DO), and analog output (AO).

8.3.4.3 Temperature sensor

The temperature sensor is a device to measure the temperature through an electrical signal, which requires a thermocouple or RTD (resistance temperature detectors). If the difference in voltage is amplified, the analog signal is generated by the device, which is directly proportional to the temperature. It is available in the dimensions of $224 \times 120 \times 60 + 40$ mm and can detect temperature in the range up to 100°C.

8.3.4.4 Smoke sensor

MQ2 gas sensor is an electronic sensor used for sensing the concentration of gases in the air, such as liquefied petroleum gas (LPG), propane, methane, hydrogen, alcohol, smoke, and carbon monoxide. MQ2 gas sensor is also known as a semiresistor. This sensor works on 5 V DC voltage and can measure concentrations of flammable gas of 300–10,000 ppm.

8.3.4.5 Servomotor

A servomotor is an electrical device that can push or rotate an object with great precision. The servomotor is used when the robot wants to rotate its robotic arms at a specific angle or distance. It is made up of a simple motor that runs through a servo mechanism. It has an average speed of 60 degrees in 0.16 seconds, a weight of 62.41 g, and a size $40.7 \times 19.7 \times 42.9$ ($L \times W \times H$) in mm.

8.3.4.6 Lithium polymer battery

Instead of a liquid battery, a lithium technology based rechargeable polymer electrolyte battery called lithium battery or lithium-ion polymer battery is used in this work (abbreviated as LiPo, LIP, Li-poly, lithium-poly). The electrolyte is formed by using a semisolid of high conductivity. These batteries provide higher specific energy than other types of lithium batteries. Due to its light weight, it is used in applications such as mobile devices, radio-controlled aircraft, and electric vehicles.

8.3.4.7 ATmega2560 microcontroller (Arduino Uno)

The ATmega2560 microcontroller (Arduino UNO) board is the brain of the robot containing the navigating and firefighting program. This microprocessor was chosen due to its speed and versatility to be used within almost any project. The versatility allows the different sensors to be controlled by one microprocessor. The processor has three analog-to-digital converter (ADC) ports, two of which have the capability of running up to 16 different channels. This would allow us to use all the analog sensors we had planned and have room for extra if needed.

8.3.4.8 DC gearmotor with wheels

A gearmotor is a specific type of electrical motor designed to produce high torque while maintaining a low horsepower, or low speed, motor output. The working voltage for a DC motor is around 5–10 V DC, while the ratio of the gear is 48:1. Suitable current for this motor is 73.2 mA. A DC motor is used to move the robot to the fire location. It has a speed of 100 rpm at 12 V and is available in the length of 46 mm and weight of 100 g. Johnson Electric provides gearmotor drive solutions to the automotive industry worldwide, including micro motors, gearboxes, and electronic interfaces.

8.3.4.9 Water pump

The water pump is an important component of this robot as it has to pump water to extinguish the fire depending on the class of the fire that occurs. The water pump selected for use in this project is small in size and lightweight. It has low noise, high effectiveness, and minimal power consumption. The optimal voltage for this water pump is 6 V, and the working voltage is around 4–12 V with a working current of 0.8 A.

8.3.4.10 Motor driver

The L298 motor driver is a high voltage, high current dual full-bridge driver designed to accept standard TTL (transistor-transistor logic) levels and drive inductive loads such as relays, solenoids, DC, and stepping motors. Two enable inputs are provided to enable or disable the device independently of the input signals. The motor driver has a dimension of $43 \times 43 \times 26$ mm with a weight of 26 g.

8.3.4.11 HC-05 bluetooth module

HC-05 has a red light emitting diode (LED), which indicates the connection status with the Bluetooth. Before connecting to the HC-05 module the red LED blinks continuously in a periodic manner. When it gets connected to any other Bluetooth device, its blinking slows down to two seconds. This module works on 3.3 V. We can connect 5 V supply voltage as well since the module has a 5–3.3 V regulator on the board. As HC-05 Bluetooth module has a 3.3 V level for receiving/transmitting (RX/TX), and the microcontroller can detect a 3.3 V level. There is no need of shifting the transmit level of the HC-05 module, but we need to shift the transmit voltage level from the microcontroller to the RX of the HC-05 module. The HC05 Bluetooth module is a universal asynchronous receiver/transmitter (UART)-serial converter module and can easily transfer the UART data through wireless Bluetooth. The Bluetooth module has a frequency of 2.4 GHz industrial, scientific, and medical (ISM) band and PIO (programmed input/output) control and comes with an integrated antenna and edge connector. The HC-05 Bluetooth module can be used in master or slave configuration.

8.4 Experimental results and discussion

The robot is aimed to detect obstacles, temperature, and smoke/flame to perform the operations resulting in the extinguishing of the fire effectively and efficiently. The design prototype of our proposed AFER is shown in Fig. 8.8.

8.4.1 Obstacle detection

As a primary objective, there must be a clear path or a direction in which the robot will move, as there should not be any obstacle restricting its movement toward the direction of the fire. So, the robot makes use of the ultrasonic sensor placed above the device, helping in detecting and making its arm operate over the obstacles and making way for the device to continue further. The robotic arm opens up the gripper. The process of obstacle detection – picking and placing the obstacle toward the left side for path clearing is shown in Fig. 8.9.

FIGURE 8.8

AFER compact prototype.

FIGURE 8.9

Detecting the obstacle and robotic arm moving toward the obstacle and opening the gripper.

8.4.2 Temperature, flame, and gas sensor detection

Similarly, after clearing its way to the destination, the next task is to detect or sense the temperature, smoke, or flame to operate its arm for sprinkling water in the direction of the fire. It makes use of the temperature sensor when it is capable of finding out the area or premises that are on fire as that would raise or increase the temperature at that particular area. The flame sensor finds out the places affected by the fire, and the robot then moves in that direction, activating the servomotor and the water pump that helps sprinkle water in that direction. The gas sensor helps in the detection of the concentration of toxic or flammable gases, which also helps the robot to judge the areas that are affected and to move in that direction. After the robot detects the fire, it extinguishes it with the water sprinkler. Fig. 8.10 shows AFER with the backup for more water supply.

Fig. 8.11 shows temperature sensor detection. Once the fire is extinguished, the robot automatically changes its direction. If no smoke, temperature, and fire is detected, then the robot goes to sleep mode. It gets switched on if the temperature sensor detects any temperature.

8.4.3 Performance measures

The performance of the proposed AFER depends on the performance of the sensor. The speed and accuracy of the AFER depend on the sensors. The input and the expected and actual output are tabulated in Table 8.1 for the flame sensor, Table 8.2 for the ultrasonic sensor, Table 8.3 for the smoke sensor, and Table 8.4 for the servomotor. It is observed from Table 8.1 that the flame sensor detects fire when a piece of paper is put up with a fire. The flame sensor is expected to detect flames in the 760–1100 nm wavelength range. The sensor is particularly sensitive to the flame spectrum. The sensor has a detection angle of 60 degrees and detects fire within the range of 100 cm with an accuracy of 95%.

It is observed from Table 8.2 that the ultrasonic sensor detects an obstacle in the path of the robot by sending electromagnetic waves in the range of 300 kHz with a maximum distance of 30 cm. However, it rotates 360 degrees after a particular range is detected with an accuracy of 95%.

FIGURE 8.10

AFER with backup for more water supply.

FIGURE 8.11

AFER detecting fire.

Table 8.1 Performance measures for flame detection (using a flame sensor).

Parameters	Flame detection performance
Sample input	Place a piece of burning paper
Expected output	Can detect flames in the 760–1100 nm wavelength range; detection angle is 60 degrees.
Actual output range	0.8 m or approximately 100 cm
Accuracy	Around 95%

Table 8.2 Performance measures for obstacle detection (using an ultrasonic sensor).

Parameters	Obstacle detection performance
Sample input	Place an obstacle in the direction of the robot
Expected output	Maximum of 30 cm at 300 kHz with resolution as low as 1 mm
Actual output range	8–10 cm with 360-degree rotation after detection
Accuracy	Around 95%

Table 8.3 Performance measures for smoke sensor (using a smoke detector).

Parameters	Smoke detection performance
Sample input	Burn a piece of paper to generate smoke
Expected output	Smoke detectors can be installed up to 41 ft. high and 10 ft. wide.
Actual output range	Detects gases in the concentration range of 200–10,000 ppm

Table 8.4 Performance measures for rotation of robotic arm (using servomotor).

Parameters	Robotic Arm performance
Power	5 V
Average speed	60 degrees in 0.16 s
Max. Torque	21.48 N-m
Max. Speed	3000 rpm
Max. Current	24.81 A

It is observed from Table 8.3 that the robot detects smoke when a piece of paper is burnt to generate smoke. It detects gases in the concentration range of 200–10,000 ppm.

It is observed from Table 8.4 that the robotic arm should be moved at a greater speed to remove obstacles from the path. The robotic arm is connected to a servomotor with a maximum speed of 3000 rpm. The average speed of servomotor is found to be 0.16 seconds.

8.5 Conclusion

This work proposes an intelligent and compact AFER is proposed to provide a technical solution to save firefighters from danger. The robot can evade obstacles, search for fire, and extinguish the fire. A device called NodeMCU (ESP8266) has been used in the project, which acts as an edge device to which all the sensors and motors are connected. The NodeMCU on the robot acts as a server that provides resources, data, and services as well as access to a Wi-Fi network from another device. The mobile or laptop through which the user can control the robot using a Wi-Fi network acts as the station. The information is sent by the station as a mode of client and server architecture. The remote user can control the movements of the robot through the information that is stored in the edge server. The designed robot is equipped with four different types of sensors to increase its capacity for handling different environments compared with QRob, which has two sensors. Temperature, flame, smoke, and ultrasonic sensors detect the corresponding parameters in the affected environment, which are processed by the controller. The body of the robot is made of a heat resistive acrylic plate to protect the electronic circuit in the robot. Its compact size enables it to enter narrow spaces. Once the location of the fire is identified, the robot stops at a certain distance (40 cm) from the fire and sends the signal to Arduino. Arduino generates the command for the water sprinkler to quench the fire from the desired location. A mini camera installed is linked to the smartphone to monitor the location. The robot has wheels covered with large belts so that it can operate easily on any terrain. It also has a water backup system, which other fire extinguishing robots do not have. There is both an automatic and manual control for the robot. The proposed robot has a 360-degree rotating robotic arm, which QRob does not have. The sensors of the proposed robot can detect for longer ranges and have more accuracy than QRob. With all these advanced features, the designed AFER is also cost-effective. The placing of the sensors, however, is a very challenging task. Future innovations in the field of robotics may allow us to communicate with a collection of robots to cooperate in a mission with the ability to receive instructions on-the-fly during an operation.

References

[1] B. Swetha Sampath, Hardware based automatic fire extinguisher robot, 12th Int. Conf. Control, Autom. Syst, ICC, Jeju Island, Korea, 2012.

[2] T.N. Khoon, P. Sebastian, A.B.S. Saman, Autonomous fire fighting mobile platform, Int. Symposium Robot. Intell. Sens. 2012 (IRIS 2012), Procedia Eng. 41 (2012) 1145–1153.

[3] V. Mangayarkarasi, Remote controlled fire fighting robot, Int. J. Trend Sci. Res. Dev. (IJTSRD) 2 (5) (2018) 820–827.

[4] M. Aliff, M.I. Yusof, N. Samsiah Sani, A. Zainal, Development of fire fighting robot (QRob), Int. J. Adv. Computer Sci. Appl. 10 (1) (2019) 142–147.

[5] R. Garu, P. Ganesh, S. Sachdeva, Multi sensor based fire fighting robot, Strad. Res. 8 (5) (2021) 712–717.

[6] J.J. Craig, Introduction to Robotics, second ed., Addison-Wesley, Reading, MA, 1989.

[7] The evolution of robotics research, IEEE Robotics & Automation Magazine, 14 (1) 2007.

[8] O. Khatib, Real-time obstacle avoidance for manipulators and mobile robots, Int. J. Robot. Res. 5 (1) (1986) 90–98.

[9] Robot introduction in Human work environment. Developments, challenges and solutions, in: 2007 IEEE International Conference on Computational Cybernetics by O. Ogorodnikova.

[10] Design and implementation of common platform for small humanoid robots, in: 2013 IEEE International Conference on Mechatronics and Automation by, Ryuki Sato; Hiroaki Matsuda; Motoyuki Fujieda; Hajime Hata; Aiguo Ming.

[11] T. Sanislav, Cyber-physical systems – concept, challenges and research areas, Liviu Miclea J. Control. Eng. Appl. Inform. CEAI 14 (2) (2012) 28–33.

[12] Cyber-Physical Systems: A Confluence of Cutting Edge Technological Streams; Parthasarathy Guturu, Bharat Bhargava, CPSR review paper.

[13] L. Sha, S. Gopalakrishnan, X. Liu, Q. Wang, Cyber-physical systems: a new frontier, 28th IEEE Int. Conf. Sens. Networks, Ubiquitous, Trustworthy Comput. 11–13 (2008) 1–9.

[14] Security aspects of cyber physical systems, in: 2018 1st International Conference on Computer Applications & Information Security (ICCAIS).

[15] A. Stefanov, C.-C. Liu, Cyber-physical system security and impact analysis, IFAC Proc. 47 (3) (2014) 11 238–11 243.

[16] N. Hassan, S. Gillani, E. Ahmed, I. Yaqoob, M. Imran, The role of edge computing in internet of things, IEEE Commun. Mag. 99 (2018) 1–6.

[17] M. Liu, F.R. Yu, Y. Teng, V.C. Leung, M. Song, Distributed resource allocation in blockchain-based video streaming systems with mobile edge computing, IEEE Trans. Wirel. Commun. 18 (1) (2019) 695–708.

[18] E. Ahmed, M.H. Rehmani, Mobile edge computing: opportunities, solutions, and challenges future gen-er, Computer Syst. 70 (2017) 59–63.

[19] Software-defined system support for enabling ubiquitous mobile edge computing. The Computer Journal, 60 (10) (2017) 1443–1457. https://doi.org/10.1093/comjnl/bxx019.

[20] B. Chen, J. Wan, A. Celesti, D. Li, H. Abbas, Q. Zhang, Edge computing in IoT-based manufacturing, IEEE Commun. Mag. 56 (2018) 9.

[21] K. Kosasih, E. Merry Sartika, M. Jimmy Hasugian, danMuliady, The intelligent fire fighting tank robot, Electr. Eng. J. 1 (2010) 1.

[22] H.P. Singh, A. Mahajan, N. Sukavanam, V. Budhraja, Control of an autonomous industrial fire fighting mobile robot, DU. J. Undergrad. Res. Innov. (2015) 124–130.

[23] S.A. Deshmukh, K.A. Matte, R.A. Pandhare, Wireless fire fighting robot, Int. J. Res. Emerg. Sci. Technol. (2015).

[24] L. Arora, A. Joglekar, Cell phone controlled robot with fire detection sensors, (IJCSIT) Int. J. Computer Sci. Inf. Technol. 6 (3) (2015) 2954–2958.

[25] A. Sharma, R. Verma, S. Gupta, S.K. Bhatia, Android phone controlled robot using bluetooth, Int. J. Electron. Electr. Eng. 7 (5) (2014) 443–448. ISSN 0974-2174.

[26] P. Saravanan, Design and development of integrated semi – autonomous fire fighting mobile robot, Int. J. Eng. Sci. Innov. Technol. (IJESIT) 4 (2015) 2.

[27] P. Sonsale, R. Gawas, S. Pise, A. Kaldate, Intelligent fire extinguisher system, IOSR Journal of Computer Engineering (IOSR-JCE) 16 (1, Ver. VIII) (2014) 59–61e-ISSN: 2278-0661, p-ISSN: 2278-8727. Available from: http://www.iosrjournals.org.

[28] P.W. Aung, W.Y. Win, Remote controlled fire fighting robot, Int. J. Sci. Eng. Technol. Res. 03 (24) (2014).

[29] S.N. Kini, R. Wadekar, S. Khatade, S. Dugane, R. Jadkar, Fire fighting robot controlled using android application, Int. J. Innov. Res. Sci. Eng. Technol. 5 (5) (2016) 7431–7436.

[30] R. Malik, Fire fighting robot: an approach, Indian Streams Res. J. 2 (I) (2012) 1–7.

CHAPTER

Applicability of edge computing paradigm for Covid-19 mitigation

9

Amit Sadanand Savyanavar[1] and Vijay Ram Ghorpade[2]

[1]*School of Computer Engineering & Technology, Dr. Vishwanath Karad MIT World Peace University, Pune, Maharashtra, India* [2]*Bharati Vidyapeeth's College of Engineering, Kolhapur, Maharashtra, India*

9.1 Introduction

The Covid-19 pandemic has disrupted human life to the extent that it has forced everyone into social distancing from their kith and kin. But human beings cannot and do not live in isolation. Nevertheless, smartphones have come to our rescue to preserve our status as social animals.

As of 4 December 2021, there were 265,211,005 Covid-19 cases and 5,258,801 deaths worldwide [1]. The virus mutated into several variants as reported in the UK, South Africa, US, India, and Brazil, which made the disease more severe, resulting in a faster spread, higher mortality, and reduced vaccine effectiveness [2]. If the virus keeps on spreading at a faster pace, it will burden the already overburdened medical facilities. The shortage of hospital rooms, medical equipment, medical staff, testing kits, etc., has been observed across many countries during the pandemic.

Here we will explore ways to deal with the Covid-19 pandemic and assess different computing paradigms to deal with its prevention so that the burden on the medical fraternity and medical facilities can be reduced. Our main focus is on the applicability of the edge computing paradigm to prevent the spread of the Covid-19 pandemic and to expedite the recovery as well.

9.2 Personal computing devices

9.2.1 Smartphones

In these testing times, smartphones [3] have brought the world to our fingertips and helped us stay connected. The apps installed have the human factor associated with them, which can guide them to work more intelligently. They can track human mobility by using location sensing technologies and keeping a record of the places visited, which can then be used to predict the future location of the user and the probability of coming into the proximity of a Covid-infected person. The Aarogya Setu app [4] developed by the National Informatics Center (NIC) of the Government of India uses GPS and Bluetooth technologies of smartphones to identify the proximity of a Covid-infected person. Smartphones are equipped with useful sensors, which can be used for health monitoring and diagnosis, as shown in Table 9.1. These sensors work with the apps installed on the smartphone to gather,

Intelligent Edge Computing for Cyber Physical Applications. DOI: https://doi.org/10.1016/B978-0-323-99412-5.00011-3

Table 9.1 Smartphone sensors to monitor and detect Covid-19 symptoms [5,6].

Health issues	Smartphone sensors
HR and HRV	Camera, microphone
RR	Camera, microphone, accelerometer
Skin	Camera, temperature
Motion	Accelerometer, gyroscope, proximity sensor, GPS
PPG signal	Camera, flashlight
SpO_2	Oxygen rate
Cough	Microphone
Stress	Accelerometer, GPS, microphone

analyze, and generate metrics. The analysis performed is quite basic since the devices are resource constrained.

HR (heart rate) and HRV (heart rate variability) are estimated from the PPG (photoplethysmography) signal, which is determined from the bare skin of the face or fingertips by using the camera. The front and rear cameras were used to monitor HR and RR (respiration rate) simultaneously. The rear camera captured the PPG signal from the fingertip placed on top of it for determining HR. The movement of the chest and abdomen is detected by the front camera for RR estimation. High-resolution cameras can be used to capture images of affected skin and fed to the mobile app for the classification of skin lesions. Temperature sensors aided with the mobile app can detect body temperature with quite a good accuracy. Motion sensors are deployed for human activity monitoring and detection that can be deployed in applications like fall detection and posture monitoring. Sensors at the back of smartphones can function as a pulse oximeter to measure SpO_2 (saturation of peripheral oxygen). Microphones can be used to capture audio signals for cough detection. Sensors such as accelerometers, GPS, and microphones can be used in stress detection. All of these, along with phone use details, can be used to assess a person's mental health condition.

Smartphones use sensors to gather data and forward it to the respective apps for further processing. These devices have limited processing and storage capacity. Hence such devices individually generate less accurate results. However, their most significant advantage lies in their ability to generate the results in real time, which is quite helpful in emergencies.

Lockdowns being extended in different parts of the country have made it extremely difficult for patients, especially the elderly and physically challenged, to reach hospitals. Mobile healthcare and telemedicine are very much in vogue today. The computing power of smartphones is either unutilized or underutilized. The dearth of medical facilities in rural areas can be addressed by using smartphones in the vicinity to keep an eye on a patient's well-being. The fight against the pandemic will be weak without the awareness and education of the precautions, symptoms, diet, medication, and repercussions of the infection. The best medium for information dissemination is, again, the ubiquitous smartphone. To sum it up, the smartphone has been a blessing in disguise during these trying times for early warning and remote surveillance.

9.2.2 Wearable devices

New York's Mount Sinai Health System and Stanford University in California collaborated on a study [7] that claims that smartwatches like Apple Watch and fitness bands from Fitbit can be pivotal in Covid-19 detection among asymptomatic patients. They measure HRV and determine whether there is lesser variability to indicate if there is an infection in the body. These devices provide nonstop monitoring and expedite quicker detection of Covid-19. They are efficient for early detection compared with Covid-19 testing, which takes at least 24 hours for diagnosis. Fitbit [8] not only measures HRV but also other physiological parameters like BP (blood pressure), SpO_2, body temperature, motion, and sleep. Proxxi Technologies, Canada [9] developed a wrist-worn wearable device that can be used to maintain social distancing at the workplace. It uses Bluetooth to provide a vibration alert about the presence of another person wearing a similar wrist device within a 6-ft. distance.

The students and the faculty of MIT-World Peace University have devised Indipro, a wearable wristband hand sanitizer. It is helpful for the Covid warriors who need to go out daily for work. As an enhancement, the team is planning to add an oxygen concentrator to it with AI and IoT technologies for predictive analysis of patients' health.

Smart glasses [10] can give the clinician insights into patients' clinical data. Medical imaging using glasses is much more simplified, and it can help to reduce the time the clinician spends with the patient by performing the tasks in parallel. All the technologies available with smartphones are available in Google Glasses as well. Phillips and Accenture Technology Lab [11] partnered to let surgeons view important information such as HR, SpO_2, and BP using the Google Glass display. People using Google Glasses can extract the statistics on their screen while choosing to maintain distance from the infected people. The pandemic has been the driver for adoption of such revolutionary technologies in healthcare. There is a growing trend toward proactive treatment so that healthy people can be kept away from hospitals for as long as possible. The availability of well-equipped wearable devices [12] has made continuous monitoring and healthcare provisioning a reality.

9.3 Edge computing

Many countries have observed that home isolation is the most prudent option to prevent the infection from spreading. Edge computing provides a framework for real-time monitoring of patients in hospitals and at home. It also enhances the possibility of monitoring patients from their homes by releasing them early from the hospitals.

Sheikh Nooruddin et al. [13] suggested an Internet-of-Things enabled fall-detection system that works with all kinds of devices like smartphones, Arduino kits, Raspberry Pi, etc. As an extension to this work, the author mentions the need of the system to work incessantly even when the Internet, i.e., the cloud or the server, connection fails.

WBAN (wireless body area network) [14] is a viable solution for patient monitoring systems. RR levels, EEG, ECG, BP, temperature, body motions, blood glucose, and other critical clinical information can be collected using sensors connected to the patient's body. Sahoo et al. [15] devised a mobile-based ECG monitoring system using sensors attached to a patient's chair. Signals from sensors were sent using Bluetooth to a smartphone that raised the alarm at the time of emergency. The mobile phone has its limitations related to processing power, storage capacity, and battery capacity. Hence the medical fraternity had apprehensions regarding such less-efficient monitoring systems. However, the Covid-19 pandemic has brought these models into focus once again. The coronavirus is highly contagious, and it is anticipated that the majority of the global population will be infected in the course of time. Positive patients require isolation and constant monitoring, of which 6% [16] of them are anticipated to require ICU beds. Both developing and developed countries are facing a paucity of hospital rooms and ICU beds, putting umpteen lives at risk and increasing the death count. With the escalating number of patients, even the best physicians are finding it difficult to make accurate decisions quickly. This is where technology can step in to be a lifesaver.

Individual sensors or smartphones have limited abilities to work as an efficient monitoring, analysis, and diagnostic system. This is where the devices in the vicinity can be brought together to create a mobile grid (MG) for edge computing. MG computing paradigm uses unutilized or underutilized devices in close proximity, such as smartphones, tablets, laptops, DVRs, etc., to form an elastic pool of resources [17,18]. Authors have also referred to it as mobile cloud computing [19,20].

The advantages of MGs are:

1. Data is gathered locally and processed without transmitting data to the cloud, which consumes bandwidth and incurs higher costs.
2. Processing data locally reduces latency and provides results in real time, which is suitable for monitoring.
3. Works without an Internet connection
4. Saves on battery consumption due to noncommunication with the cloud.
5. Patient data is processed locally, hence alleviating user's privacy concerns.

Vital physiological parameters are extracted using sensors, and complex physiological models are required to evaluate and interpret the information gathered. Edge computing with MGs facilitates real-time processing of this data by using a resource allocation framework. The complex physiological models involve a lot of subtasks that can be assigned to individual devices in the MG for computation. A resource allocation framework can ensure efficient resource allocation by considering the characteristics of each device prior to task allocation [21]. Different vital indicators, such as EMG, ECG, RR, EEG, GSR, temperature, blood volume, and pulse and oxygen level, can be captured by MGs. They also execute computational models to determine hypoxia, stress, sleep quality, cognitive performance, etc.

Prabhdeep Singh et al. [22] devised a fog unit integrated with an AI module to improve the accuracy of Covid-19 prediction. The AI unit was an ensemble of three classifiers: Random forest, generative adversarial networks, and naive Bayes. IoT devices collected data and forwarded it to fog nodes for processing, storing, and generating alerts. Fog devices are resource constrained. The response time surged by almost six times when the cloud is used instead of the fog unit. Such an approach could be useful for homes, housing societies, and small hospital establishments.

9.4 Proposed work

Edge computing with MGs provides a framework for patients to be monitored in real time, especially in their homes. In such paradigms, a few challenges exist like efficient resource provisioning, failure handling, energy preservation, etc. To deal with these issues, we have proposed an efficient resource allocation model (ERAM) [17,18,23], which is a resource allocation, failure handling, and energy preserving model for edge computing with MGs.

As shown in Fig. 9.1, ERAM works in the following manner:

1. In MG, any node can act as an initiator and start the application execution.
2. The initiator runs ERAM to assign application tasks to nodes.
3. The device discovery module in ERAM is used to identify the nodes that make up an MG.
4. Using the location prediction (LocP) [24] mechanism, identified nodes determine their mobility status, i.e., their grid node stability. LocP will assist in identifying if a node is stable or unstable, and each node will inform the ERAM algorithm of this condition.
5. ERAM would then use AMT (application metadata template), which specifies the details of all the subtasks involved in the application. It mentions the minimum number of nodes necessary for particular task execution and whether the task is computation bound or communication bound.
6. Following that, the node classifier module is used. It employs a rough set theory based node classification method to divide nodes into three categories: Positive, boundary, and negative, which helps in resource allocation. The positive set comprises the nodes that are definitely used for task execution. The negative set enlists the nodes that are discarded from executing the application tasks. The boundary set comprises the nodes that can possibly be considered for task allocation if there is a scarcity of nodes for computation. This classifier gathers node

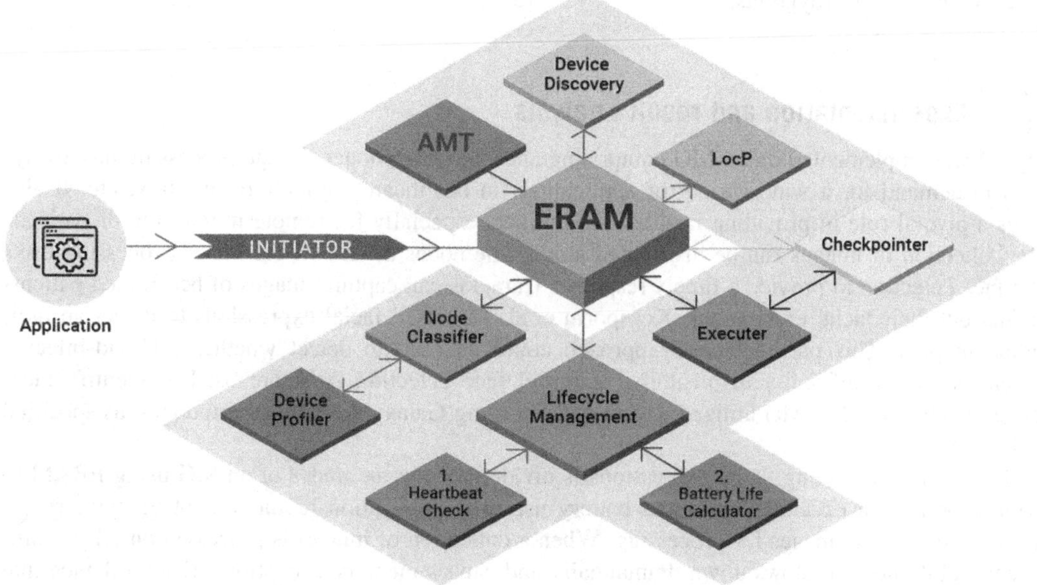

FIGURE 9.1

ERAM system components.

information such as processing power, residual battery, connectivity, available RAM, and storage space for all nodes in the MG, using device profiler. These attributes are referred to as condition attributes. When the MG's collective battery usage surpasses a predetermined threshold, ERAM activates the node classifier. The three categories of nodes are determined using the rough set theory, as mentioned below. Nodes are referred to as objects represented as "o" in the rough set theory.

1. Find the R-indiscernibility relation that consists of equivalence classes $[x]_R$
   $R(C) = \{(oi, oj) \in U \quad : \forall a \in A, f(oi, \quad a) = f(oj, \quad a) \quad \}$
2. Determine the set of objects whose decision attribute $d = Yes$
   $Cyes = \{o| \quad \forall o \in U, d = Yes\}$
3. Derive the lower approximation of C_{yes}
   $\underline{R}Cyes = \{o \in U \quad :[o]R \quad \subseteq Cyes \quad \}$
4. Derive the upper approximation of C_{yes}
   $\overline{R}Cyes = \{o \in U \quad :[o]R \quad \cap Cyes \neq \phi \quad \}$
5. Boundary region includes those objects that are possibly, but not certainly, part of $d = Yes$ membership class. Boundary region of C_{yes} is
   $BND = \overline{R}Cyes - \quad \underline{R}Cyes$
   Lower approximation represents the positive set, whereas the nodes not part of the upper approximation belong to the negative set.
7. ERAM deploys the appropriate MG nodes as either executer or checkpointer nodes.
8. The lifecycle management module is used to continually check the presence and involvement of each node in the MG. It checks if a node is alive or dead and determines the battery level of each node using waypoints.

9.4.1 Experimentation and result analysis

ERAM was implemented on an MG comprising android smartphones. Image processing has always been in demand for a wide variety of applications in healthcare, agriculture, military, etc. It also plays a pivotal role in providing healthcare solutions, especially for remote monitoring of patients. The collection of images can be distributed among the nodes of an MG for image processing tasks like face detection to provide a timely response. Cameras can capture images of bed-ridden patients to find out their facial expressions. Kornporm et al. [25] used facial expressions to detect an individual in pain. This pain detection approach could be used to detect whether a Covid-infected patient is having difficulty in breathing. Android face detection APIs are used to identify face-related landmarks. The MG utilized Redmi and Samsung Galaxy Android smartphones, as specified in Table 9.2.

The work of detecting facial expressions is divided among the nodes of an MG using ERAM to reduce application execution time and battery use. The application requires a lot of memory for storing and loading images for processing. When a collection of images is processed on a less competent mobile node, it slows down dramatically and causes memory exceptions. Even a device like this can execute resource-intensive tasks by offloading workload to more efficient execution nodes. The input required for this application is huge, yet the result delivered is of trivial size. Here 10 distinct image files were used, with the same files replicated to create a workload of 50 files.

Table 9.2 MG node specifications.

Nodes in MG	Node details
1	Redmi 2
2	Redmi 2, Galaxy OnNxt
3	Redmi 2, Galaxy OnNxt, Galaxy E7
4	Redmi 2, Galaxy OnNxt, Galaxy E7, Galaxy Grand 2
5	Redmi 2, Galaxy OnNxt, Galaxy E7, Galaxy Grand 2, Galaxy On7 pro
6	Redmi 2, Galaxy OnNxt, Galaxy E7, Galaxy Grand 2, Galaxy On7 pro, Galaxy On5
7	Redmi 2, Galaxy OnNxt, Galaxy E7, Galaxy Grand 2, Galaxy On7 pro, Galaxy On5, Galaxy On8
8	Redmi 2, Galaxy OnNxt, Galaxy E7, Galaxy Grand 2, Galaxy On7 pro, Galaxy On5, Galaxy On8, Lenovo Tab M10
9	Redmi 2, Galaxy OnNxt, Galaxy E7, Galaxy Grand 2, Galaxy On7 pro, Galaxy On5, Galaxy On8, Lenovo Tab M10, Galaxy A9

9.4.1.1 Effect of varying workload

The workload and the number of executer nodes were varied as part of the experiment. Fig. 9.2 shows that for facial expression detection, 10–30 picture files were used and the number of executer nodes in MG varied from 1 to 7. It can be seen that with the increase in the number of nodes, the application completion time does not decrease considerably. This is because, as the number of connections increased, the initiator's overhead for managing and maintaining concurrent connections increased. In addition, as the number of connections grows, so does the time taken to distribute the workload. For this scenario, the minimal aggregate time for application execution was provided by 4–5 executer nodes. Fig. 9.2 illustrates the application execution time when the image files were increased to 20. When 4 executer nodes were used instead of only one, the time needed was decreased by nearly 30%. As the number of executer nodes increased from 5 to 7, the time required increased. Only 10% of the time is saved for 7 nodes.

More data must be transferred to the nodes when the number of files was increased to 40. As seen in Fig. 9.3, the highest reduction in completion time is about 18% for four nodes. Additional data must be transmitted to the nodes if the number of files grows to 50. Consequently, the degree of computing required has grown significantly. As a consequence, the application completion time for 4 nodes compared with a single node is reduced by 27%.

9.4.1.2 Effect on application completion time by varying workload and number of nodes

The effect of varying the workload and the number of nodes are combined in this experiment. Fig. 9.4 shows the results as the number of files increases from 10 to 50 and the number of MG nodes increase from 4 to 5. It represents the fact that time savings are proportional to the scale of the workload, i.e., the number of files and the computational devices. In comparison to computation using four nodes, a workload of 50 files and five nodes saves 21% time.

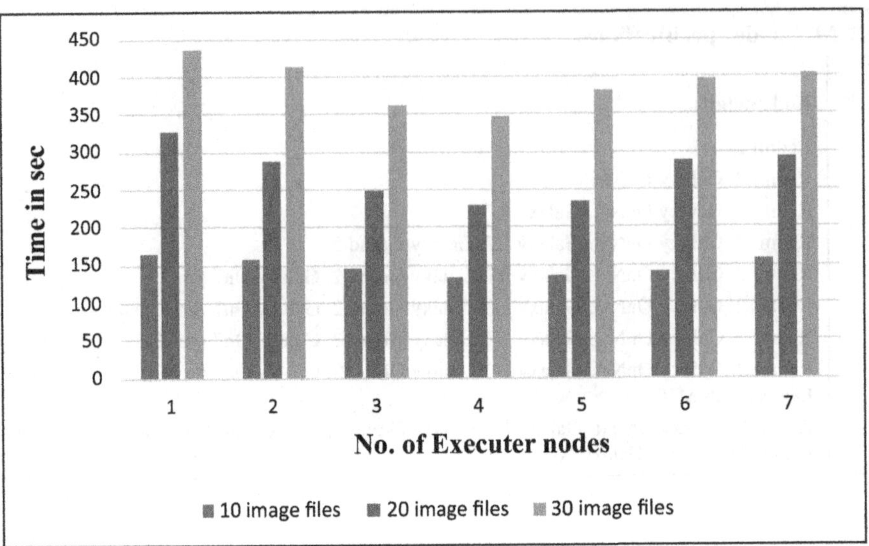

FIGURE 9.2

Number of executer nodes versus application completion time.

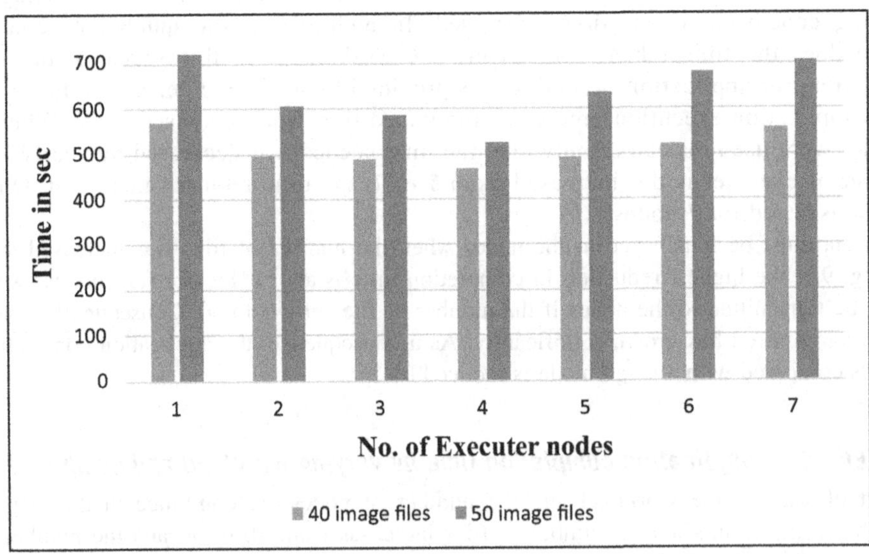

FIGURE 9.3

Number of executer nodes versus application completion time (40–50 files).

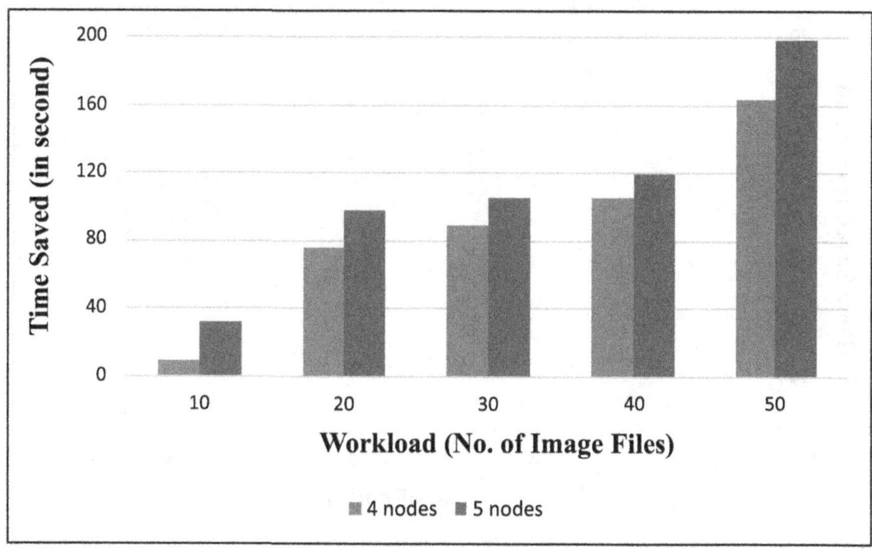

FIGURE 9.4

Workload versus time saved.

9.4.1.3 Effect on application completion time by varying number of nodes

Fig. 9.5 demonstrates how changing the nodes in MG from 2 to 9 saves time for application execution for a workload of 50 files. When the number of nodes is increased from 2 to 5, the application execution time is reduced by nearly half. When the number of nodes is increased from 5 to 9, the application time is reduced by 18%. This is because the number of connections that need to be maintained and managed has been increased. Additionally, the time needed for the distribution and transmission of files across multiple connections has also increased.

9.4.1.4 Effect of number of executer nodes on average battery usage

The aim of ERAM is not only to expedite the application completion time as desired in emergency scenarios but also to enhance the battery life of the nodes in the MG. ERAM does this by dividing the workload across the competent nodes and distributing the battery drain across the nodes. Fig. 9.6 shows that using 7 executer nodes for 50 files resulted in a maximum average energy savings of 64%. The average battery usage for the MG nodes is negligible for 10 files. As the workload grows, a significant amount of energy is conserved.

So we can conclude that ERAM provides an efficient resource provisioning model for edge computing, capable of expediting the computations and monitoring the health condition of patients in real time. It also aids in reducing the battery consumption for the computing devices. It uses the idle devices in the patient's vicinity without burdening a single computing device.

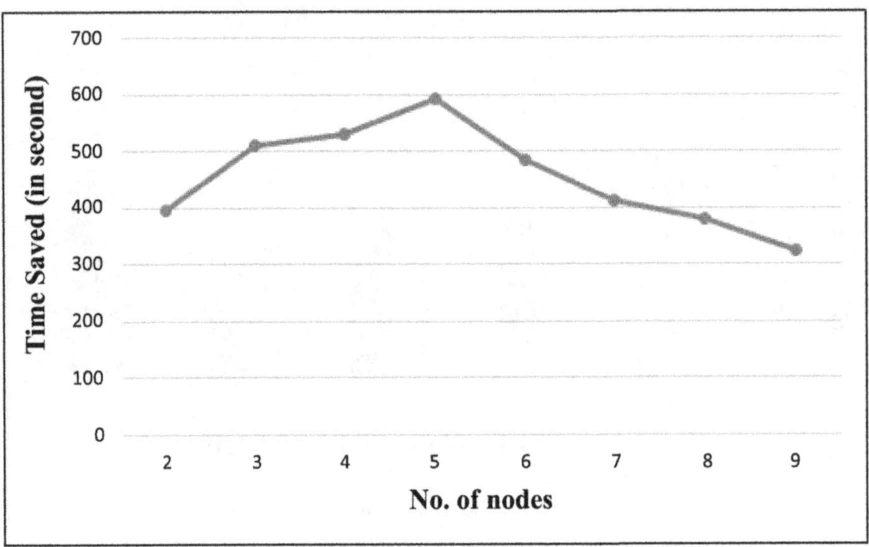

FIGURE 9.5

Number of nodes versus time saved.

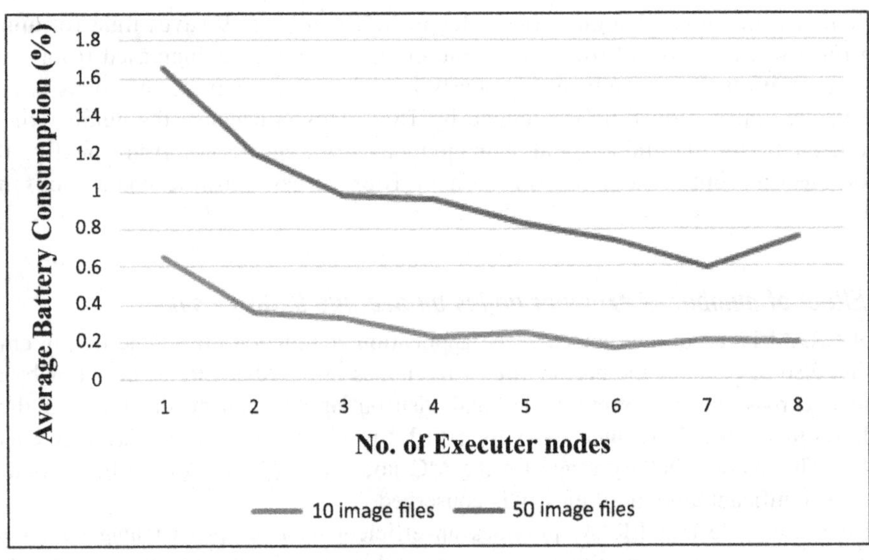

FIGURE 9.6

Number of executer nodes versus average battery consumption (%).

9.5 Cloud computing

Cloud computing (CC) provides a state-of-the-art option for remote monitoring of patients since it adds many capabilities and alleviates many limitations of personal and MG computing systems. CC provides three types of services: Infrastructure as a Service (IaaS), e.g., Amazon storage service S3; platform as a service (PaaS), providing both storage and service, e.g., Google App; and software as a service (SaaS). The quality of service (QoS) for remote monitoring systems can be enhanced by adding CC to wireless sensor networks. It facilitates access to patient data from anywhere, at any time, and in real time. Features such as high-volume data processing, analysis, and data sharing at low cost reduce the overhead cost of data management for the hospitals.

Aarogya Setu is one of the best examples of the use of CC for Covid-19 prevention. It is a contact tracing app used to determine if a person has been within a 6-ft. distance of an infected person. It scans the database server of all the positive cases in India and creates a social contact graph. The person can then be alerted to seek medical help, test, quarantine, etc. As an extension, a prediction module can be incorporated to predict the possibility of being infected by the virus based on a person's mobility history. Several Covid-19 contact tracing applications have been deployed throughout the world to detect and restrict the spread of infection.

Machine learning (ML) and deep learning (DL) techniques can play a vital role in the prediction and detection of the infection. These techniques improve the accuracy of prediction and detection when deployed on the cloud. CC infrastructure facilitates the colossal storage and big data processing required by these techniques.

9.5.1 Machine learning

Automated Covid-19 prediction systems are feasible today using ML. Villavicencio et al. [26] applied five supervised ML algorithms namely, random forest, J48 decision tree, naive Bayes, K-nearest neighbors, and support vector machines (SVMs) for Covid-19 prediction. The ML algorithms are efficient and accurate for disease predictions. SVMs showed an accuracy of 98.81% for Covid-19 prediction. Random forest showed similar accuracy but with greater mean square error. The authors used ML software WEKA for the analysis and Covid-19 dataset from Kaggle.

A hybrid fog-cloud framework [27] can be designed which works at two levels. In the first level the fog unit is responsible for time-constrained symptom based computing and prediction of the disease. The cloud is used at the second level for batch processing of collected dataset for categorizing patient as normal, Covid patient, or pneumonia patient.

ML techniques like SVMs, Artificial Neural Network (ANN), and random forest were deployed by Alotaibi et al. [28] for predicting the severity of the patients. Random forest gave the best accuracy of 90.83%, followed by SVM and ANN. Ghosal et al. [29] used a linear regression model to forecast the number of fatalities in India in the present situation. Yue et al. [30] extracted characteristics from 72 pneumonia lesions associated with 31 patients using ML approaches such as random forest and logistic regression. Sensitivity of 1.0 and specificity of 0.89 were obtained from logistic regression, whereas the random forest model has a specificity of 1.0 and a sensitivity of 0.75.

Patrikar et al. [31] adapted the SEIR framework to examine the influence of social distancing on the transmission of coronavirus. The findings of the experiment confirmed that social distancing can minimize the infection by 78%.

9.5.2 Deep learning with radiological images

Nora El-Rashidy et al. aggregated X-ray images of the chest which were stored on the cloud. These images were used to predict the existence of corona virus infection. DL technique CNN takes images as input to the image processing module and assigns weights to different characteristics in the image. These weights are then utilized to determine whether or not the patient is infected. Hemdan et al. [32] created COVIDX-Net, a diagnostic framework that employs seven DL models to analyze X-ray images. The accuracy of diagnosis was almost 90%.

Wang et al. [33] proposed the COVID-Net model, which uses a deep convolutional neural network (CNN) to differentiate between pneumonia and Covid-19 cases. The model utilized X-ray scans to determine the severity of the chest infection. Over 14,000 X-ray scans were tested, with a recognition accuracy of 93.3%. On similar lines, Apostolopoulos et al. [34] deployed CNN with transfer learning as a classification approach for COVID-19 cases. Over 1427 X-ray scans were tested and the accuracy of recognition was 98.75%.

ML and DL techniques could be deployed on Cloud for accurate prediction and detection of Covid-19 infection. Pham et al. conceived the idea of CoSHE [35] (cloud based smart home environment) for providing remote healthcare solutions to the patients at their homes. Data gathered using sensors was processed at a cloud to acquire comprehensive information about the patient. The sole purpose of the system was to provide efficient healthcare solutions at home.

Debauche et al. [36] proposed IoT fog with Cloud-based health monitoring model. The approach was focussed on improving the quality of healthcare facility for the patients at home by

Table 9.3 Comparison of computing paradigms.

	PC	EC with MG	CC
Available with	Individuals	Families/Small organizations	Large organizations
Availability	Very high	Quite high	High
Response Time	Low (in milliseconds)	Medium (in milliseconds)	High (in seconds)
Accuracy	Low	Moderate	High
Computational complexity	Lightweight	Moderate	Very high
Cost	Low	Medium	Very high
Delay	Low	Moderate	High
Internet	Not necessary	Not necessary	Mandatory
Privacy issues	None	Low	High
Big data analysis	Not possible	Not possible	Possible
Storage capacity	Low	Medium	Very high
Energy consumption	Low	Medium	High
Task complexity	Low	Medium	High

integrating Fog and Cloud services. Patients' daily activities were monitored by using physiological and environmental signals. Proposed system had three layers comprising of sensors, fog and cloud.

Comparison of the discussed computing paradigms is shown in Table 9.3.

9.6 Conclusion

The widespread availability of smartphones and various other personal computing devices across the population has been a blessing in disguise during the Covid-19 pandemic. It enabled remote, home-based healthcare solutions for patient monitoring, diagnosis, and timely medical advice. Personal computing paradigm provides timely but very basic information of the patient's health condition. Edge computing with mobile grids provides more computing capacity for deploying complex models for monitoring the health condition of an infected patient. ERAM enables resource provisioning for computing in the vicinity of the data source itself, enabling real-time generation of results to save human life. Cloud computing is the most sophisticated and advanced option but with quite a significant delay. ML and DL techniques can be executed on a cloud to improve the accuracy of these systems. Nevertheless, each of these paradigms can play a significant role in dealing with any such highly transmissible infections as COVID-19 in the future as well.

Author's contributions

Dr. Vijay Ghorpade was involved in manuscript conceptualization; Amit Savyanavar was involved in literature review, experimentation, and writing of manuscript

Disclosure

None.

Funding

No funding was involved in this study.

Ethical approval

None required.

Patient consent for publication

None required.

Declaration of competing interest

The authors declare No conflict of interest.

References

[1] Worldometer. COVID Live Update, 4th December 2021. https://www.worldometers.info/coronavirus/

[2] Centers for Disease Control and Prevention (CDC). SARS-Cov-2 Variant Classifications and Definitions, 17 May 2021. Available online: https://www.cdc.gov/coronavirus/2019-ncov/cases-updates/variant-surveillance/variant-info.html (accessed on 23 May 2021).

[3] K. Iyengar, G.K. Upadhyaya, R. Vaishya, V. Jain, COVID-19 and applications of smartphone technology in the current pandemic, Diabetes Metab. Syndr. 14 (5) (2020) 733–737. Available from: https://doi.org/10.1016/j.dsx.2020.05.033.

[4] K.P. Iyengar, R. Jain, D.A. Samy, V.K. Jain, R. Vaishya, K. Suraliwala, Exploring the role and mechanism of COVID-19 apps in fighting the current pandemic, Indian J. Med. Sci. 73 (1) (2021) 36–40.

[5] S. Majumder, M. Jamal Deen, Smartphone sensors for health monitoring and diagnosis, Sensors 19 (2019) 2164. 103390/s19092164.

[6] S. Nizar Al Bassama, A. Asif Hussain, J. Al Qaraghuli, E.P. Khan, V.L. Sumesh, IoT based wearable device to monitor the signs of quarantined remote patients of COVID-19 (2021). Available from: https://doi.org/10.1016/j.imu.2021.100588.

[7] https://www.news18.com/news/tech/apple-watch-garmin-fitbit-trackers-can-detect-covid-19-cases-days-in-advance-finds-study-3299159.html

[8] V. Kumar, D. Singh, M. Kaur, R. Damasevicius, Overview of current state of research on the application of artificial intelligence techniques for COVID-19, PeerJ Comput. Sci. (2021). Available from: https://doi.org/10.7717/peerj-cs.564.

[9] C. Russey, Proxxi launches halo wearable wristband to maintain social distance at work during Covid-19 pandemic, Health & Wellness (2020). Available from: https://www.wearable-technologies.com/2020/04/proxxi-launches-halo-wearable-wristband-to-maintain-social-distance-at-work-during-covid-19-pandemic/.

[10] T. Munusamy, R. Karuppiah, N. Bahuri, S. Sockalingam, C.Y. Cham, V. Waran, Telemedicine via smart glasses in critical care of the neurosurgical patient-COVID-19 pandemic preparedness and response in neurosurgery, World Neurosurg. 145 (2021) e53–e60. Available from: https://doi.org/10.1016/j.wneu.2020.09.076.

[11] https://www.healthcare.digital/single-post/2017/08/23/7-ways-google-glass-is-revolutionising-healthcare

[12] F.A. Kraemer, A.E. Braten, N. Tamkittikhun, D. Palma, Fog computing in healthcare: a review and discussion, IEEE Access. (2017). Available from: https://doi.org/10.1109/ACCESS.2017.2704100.

[13] S. Nooruddin, Md Milon Islam, F.A. Sharna, An IoT based device-type invariant fall detection system, Internet Things 9 (2020) 100130. Available from: https://doi.org/10.1016/j.iot.2019.100130. ISSN 2542-6605.

[14] A. Manirabona, L.C. Fourati, A 4-tiers architecture for mobile WBAN based health remote monitoring system, Wirel. Netw. 24 (2017) 2179–2190.

[15] P.K. Sahoo, H.K. Thakkar, M.-Y. Lee, A cardiac earlywarning system with multi channel SCG and ECG monitoring for mobile health, Sensors 17 (2017) 711.

[16] World Health Organization, Operational considerations for case management of COVID-19 in health facility and community. Interim guidance, Pediatr. Med. Rodz. 16 (2020) 27–32.

[17] A.S. Savyanavar, V.R. Ghorpade, Efficient resource allocation scheme for on-the-fly computing based mobile grids, Int. J. Inf. Technol. (2018). Available from: https://doi.org/10.1007/s41870-018-0269-y.

[18] Amit S. Savyanavar, Vijay R. Ghorpade, Application checkpointing technique for self-healing from failures in mobile grid computing, Int. J. Grid High. Perform. Comput. 11 (2) (2019).

[19] S. Nora El-Rashidy, S.M. El-Sappagh, H.M. Riazul Islam, S.A. El-Bakry, End-to-end deep learning framework for coronavirus (COVID-19) detection and monitoring, Electronics 9 (2020) 1439. Available from: https://doi.org/10.3390/electronics9091439.

[20] H. Viswanathan, E.K. Lee, I. Rodero, D. Pompili, Uncertainty-aware autonomic resource provisioning for mobile cloud computing, IEEE Trans. Parallel Distrib. Syst. 26 (8) (2015) 2363–2372. Available from: https://doi.org/10.1109/TPDS.2014.2345057.

[21] H. Viswanathan, B. Chen, D. Pompili, Research challenges in computation, communication, and context awareness for ubiquitous healthcare, IEEE Commun. 50 (5) (2012) 92–99.

[22] P. Singh, R. Kaur, An integrated fog and Artificial Intelligence smart health framework to predict and prevent COVID-19, Global Transit. 2 (2020) 283–292. Available from: https://doi.org/10.1016/j.glt.2020.11.002. ISSN 2589-7918.

[23] A.S. Savyanavar, V.R. Ghorpade, Node classification model for on-the-fly computing based mobile grids using rough set theory, Int. J. Inf. Technol. Secur. 13 (3) (2021) 15–25.

[24] A.S. Savyanavar, V.R. Ghorpade, Node mobility prediction in mobile grid computing, Int. J. Emerg. Trends Technol. 2 (2) (2015) 342–347. ISSN: 2455-0124.

[25] K. Pikulkaew, W. Boonchieng, E. Boonchieng, V. Chouvatut, 2D facial expression and movement of motion for pain identification with deep learning methods, IEEE Access. 9 (2021) 109903–109914.

[26] C.N. Villavicencio, J.J.E. Macrohon, X.A. Inbaraj, J.-H. Jeng, J.-G. Hsieh, COVID-19 prediction applying supervised machine learning algorithms with comparative analysis using WEKA, Algorithms 14 (2021) 201. Available from: https://doi.org/10.3390/a14070201.

[27] A. Kallel, M. Rekik, M. Khemakhem, Hybrid-based framework for COVID-19 prediction via federated machine learning models (version 1), TechRxiv (2021). Available from: https://doi.org/10.36227/techrxiv.13626755.v1.

[28] A. Alotaibi, M. Shiblee, A. Alshahrani, Prediction of severity of COVID-19-infected patients using machine learning techniques, Computers 10 (3) (2021) 31. Available from: https://doi.org/10.3390/computers10030031.

[29] S. Ghosal, S. Sengupta, M. Majumder, B. Sinha, Linear regression analysis to predict the number of deaths in India due to SARS-CoV-2 at 6 weeks from day 0 (100 cases—March14th 2020), Diabetes Metab. Syndrome: Clin. Res. Rev. 14 (4) (2020) 311–315. Available from: https://doi.org/10.1016/j.dsx.2020.03.017.

[30] H. Yue, Q. Yu, C. Liu, Y. Huang, Z. Jiang, C. Shao, et al., Machine learning-based CT radiomics method for predicting hospital stay in patients with pneumonia associated with SARS-CoV-2 infection: a multicenter study, Ann. Transl. Med. 8 (14) (2020) 859. Available from: https://doi.org/10.21037/atm-20-3026.

[31] S. Patrikar, D. Poojary, D.R. Basannar, D.S. Faujdar, R. Kunte, Projections for novel coronavirus (COVID-19) and evaluation of epidemic response strategies for India, Med. J. Armed Forces India 76 (3) (2020) 268–275. Available from: https://doi.org/10.1016/j.mjafi.2020.05.001.

[32] E.E.D. Hemdan, M.A. Shouman, M.E. Karar, COVIDX-Net: a framework of deep learning classifiers to diagnose COVID-19 in X-ray images (2020). Available from: http://arxiv.org/abs/2003.11055.
[33] L. Wang, Z.Q. Lin, A. Wong, COVID-Net: a tailored deep convolutional neural network design for detection of COVID-19 cases from Chest X-ray images, Sci. Rep. 10 (1) (2020) 19549. Available from: https://doi.org/10.1038/s41598-020-76550-z.
[34] I.D. Apostolopoulos, T. Bessiana, COVID-19: automatic detection from X-ray images utilizing transfer learning with convolutional neural networks, Phys. Eng. Sci. Med. 43 (2) (2020) 635–649. Available from: https://doi.org/10.1007/s13246-020-00865-4.
[35] M. Pham, Y. Mengistu, H. Do, W. Sheng, Delivering home healthcare through a cloud-based smart home environment (CoSHE), Future Gener. Comput. Syst. 81 (2018) 129–140.
[36] O. Debauche, S. Mahmoudi, P. Manneback, A. Assila, Fog IoT for health: a new architecture for patients and elderly monitoring, Procedia Comput. Sci. 160 (2019) 289–297.

CHAPTER

Design and implementation of UWB-based cyber-physical system for indoor localization in an industry environment

10

Shilpa Shyam[1], Sujitha Juliet Devaraj[2], Kirubakaran Ezra[2], Jeremy Delattre[3] and Geo Kingsly Lynus[4]

[1]Department of Computer Science and Engineering, Karunya Institute of Technology and Sciences, Coimbatore, Tamil Nadu, India [2]Department of Computer Science and Engineering, Karunya Institute of Technology and Sciences, Coimbatore, Tamil Nadu, India [3]Centre of Excellence-Interiors, Alstom, Valence, France [4]Department of Design to Cost, Alstom Transport, Bengaluru, Karnataka, India

10.1 Introduction

Systems that tend to coordinate the computational aspects with the physical world are termed cyber physical systems. A communication and a computing layer look over the functions of such systems. The interactive relationship between the two layers is of vital importance. The world is witnessing a constant change with such cyber physical systems among which robotic surgery, intelligent buildings, and smart factories are mere examples. Cyber physical systems are admitted to be the core of the industry 4.0 [1–3] revolution. They are capable of recording and affecting data using various physical components. They inherit the capability of evaluating such recorded data with the physical world. The result of such interactive and capable systems leads to eye-dropping innovations every single day.

In the eyes of the manufacturing sector, indoor-based cyber physical systems are internet-based structures that are vastly used for monitoring, tracking, and navigating, contributing majorly to industry 4.0. Such indoor-based systems have extended their capabilities to smart healthcare, robotics, smart buildings, and even in precision agriculture. Indoor positioning systems are the core applications of cyber-physical systems. The inability of global positioning systems indoors has led to the discovery of indoor positioning systems. Indoor navigation systems can be distinguished as wearable devices [4], wall-mounted devices [5], or intelligent systems [6] working together to provide accurate and precise location details. The devices constitute a system module that locates the position of the object, a module that helps the user in navigating from a precise location to the tag location, and a module that helps in creating an interaction between the first and the second module [7].

Cyber-physical systems involved in indoor navigation are currently witnessing huge demand Starting from the aviation industry [8] to the healthcare industry [9], cyber physical systems based

Intelligent Edge Computing for Cyber Physical Applications. DOI: https://doi.org/10.1016/B978-0-323-99412-5.00010-1

on indoor positioning have become very prominent as intelligent systems. A variety of such systems are used in the aviation industry to track and monitor customer and staff valuables [10]. In the Sydney airport, Apple Maps are used for the navigation of customers and staff in certain terminals of the airport. The health industry makes use of cyber-physical systems for monitoring and tracking patient data [11]. Other sectors, including e-commerce and the logistics industry, have also started using cyber-physical systems for various applications. The industry giant Google put forth a tango augmented reality system that promises to find the user location precisely just by using the mobile device.

10.1.1 Techniques and technologies in cyber physical-based indoor positioning systems

All cyber-physical-based indoor positioning systems work on a common principle [12]. They calculate the distance and range between two given devices accurately, which is executed in two ways: The first one uses received signal strength for measuring, while the other one estimates the time-of-flight measurement from several devices. Conventional methods of tracking and monitoring are done by Wi-Fi, Bluetooth, RFID, Dead Reckoning, Ultrasonic, and ZigBee [13–16] RFID makes use of radio waves for identifying an object that tends to change frequencies during identification. RFID tracking systems are implemented for identifying dynamic objects with less than 1 m accuracy [17], proving its capability in applications where tracking is highly necessary. For tracking and monitoring in short distances, Bluetooth technology is most sought after. Cyber-physical systems can be highly expensive, And as a relief to such cost measures, ZigBee can be used as it creates mostly low-cost systems using low power [18]. Cyber physical-based indoor positioning systems are much sought after for smart homes and smart buildings. More recently, the ultra-wideband (UWB) technology has gained much prominence, particularly when industry requirement is asked for, due to its increased accuracy and precision capability. [19].

Indoor positioning techniques include triangulation, proximity, fingerprinting, and vision analysis. When object identification is made using geometric triangle calculations, it is known as triangulation, which is further segregated as angulation and lateration. When distance alone is used as a measure, it is known as lateration, while angulation measures both angles and distances. When object identification is done using a known reference point, it is known as proximity analysis but involves a huge number of detector identification. The selection of proper technique and technology is highly recommended while implementing a cyber-physical-based indoor positioning system. The authors have put their effort into selecting the right technique and technology to bring about a UWB-based cyber-physical system with an edge device. It implements the exact positioning of objects indoors, using the technique and technology of TDoA (time difference of arrival) and triangulation.

10.2 Role of UWB in industry 4.0

A train of impulses is used by the UWB technology to communicate information for precise and accurate object identification. Considering its various advantages, UWB is most sought after for industrial applications. Research is still going on for high-end advanced UWB systems for smart applications. Literature proves that there are minimal-cyber physical-based accurate UWB systems [20–28]. The role of UWB is significantly felt in the smartphone [29] industry, industrial applications, and various other applications. In Shyam et al. [30] the tracking of vehicles in a parking plot using UWB is explained. The architecture promises results that would help in contributing to a smart city with the help of UWB applications. A complete analysis on the setup of a UWB-based system for industries has also been done [11].

The design and architecture proposed exhibit a low system with high accuracy. A higher accuracy system for indoors by merging IMU and UWB with the use of the extended Kalman filter has been found to have a higher accuracy in cases of 2D and 3D scenarios [31]. A collation done between ultrasonic and UWB localization capability proved that UWB has its own advantages for the industry environment [32].

Table 10.1 [33] shows a comparison between the currently available UWB-based system in the market. (TDoA: time difference of arrival; AoA: angle of arrival.)

10.2.1 Motive behind this work

The motive behind this work was raised due to the factors discussed in the following sections.

10.2.1.1 Proliferation of industrial cyber-physical systems

Considering a leap in the production of industrial-based cyber-physical systems and the complexities involved, the invention of this system has been brought forth. The molding of edge devices and logistics promises to come up with a solution and escalates asset tracking systems for industries.

10.2.1.2 Goal of an upgraded smart factory

Tracing and monitoring static and dynamic objects continuously on a huge work floor is a monotonous and tedious job. It requires huge labor and time as well. Observing such disadvantages, this work pushes toward smart production and ensures that it is attractive and sustainable at the same time.

10.2.1.3 Inclusion of logistics

The real-time tracking of objects and updating them in a certain database is an inevitable step in creating a timeline of the position of the traced objects. The in and out movement of assets intended would be traced and updated for future purposes.

Table 10.1 Comparison of present UWB systems.

Metrics	Decawave	Sewio	Ubisense
Frequency	5–8 GHz	5–9 GHz	6–10 GHz
Bandwidth	500,800 MHz	500 MHz	800 MHz
Technique implemented	TDoA	TDoA	TDoA + AoA
Measured accuracy	<25 cm	Less than 25 cm	Less than 10 cm
Coverage	280 m	45 m	60 m

10.2.2 Problem formulation

An infinite number of objects or assets is required to be traced in a huge warehouse or a manufacturing site. When tracing and monitoring is done by humans, results are not found to be accurate and precise. Hence a system that would be able to detect and provide the exact location of the object is highly recommended. Several authors have come up with various solutions intending to solve the problem using various indoor positioning technologies, but the accuracy and precision they provide vary. Moreover, a 3D implementation of the traced objects is hardly observed. Keeping these limitations in mind the authors have decided to formulate an antidote using UWB technology. The proposed work is concerned not only with the positioning of assets but also with the monitoring and continuous observation using a novel software application.

10.2.3 Objective behind this work

The proposed system promises to accomplish the following:

- Tracking of assets: The position of assets would be traced using UWB technology along with triangulation techniques using TDoA.
- Navigation of assets: A software application would help the user to navigate toward the assets from a common reference point in the industry once the position is determined.
- Monitoring of assets: The history and status of assets inside the industry would be constantly monitored using data analysis software.

10.3 Methodology

10.3.1 Tracking of assets

Assets can be traced on a huge work floor with the proposed simplified architecture (Fig. 10.1).

10.3.1.1 Simplified architecture

The proposed architecture shows the transfer of information of tags from anchors to the end user. The tags are described as objects or assets that require to be traced. The anchors are wall fixed and have wide coverage of their own, and are kept at a minimal distance depending on the place of implementation. The anchors are initially clock synchronized. The system works on TDoA [34,35]. The tags on sending a pulse of 1 Hz during regular intervals are received by the anchors. The anchors located nearest to the tag receive the pulse at the earliest, and the timing is recorded. A minimum of four anchors closest to the tag is considered. The communication protocol followed here is Transmission Control Protocol/Internet Protocol (TCP/IP). The information about the tags is carried by the anchors to the 24 port PoE (power over Ethernet) switch using a PoE cable. The PoE cable helps in transferring data without the need for extra cables for power. The information is passed on to the Raspberry Pi, which acts as the edge device in this case. The algorithm calculation is done with the edge device. Considering the vast location of an industry, a router is employed for the end results to be shared between the user and the edge device. Users can receive the tag information using any smart devices.

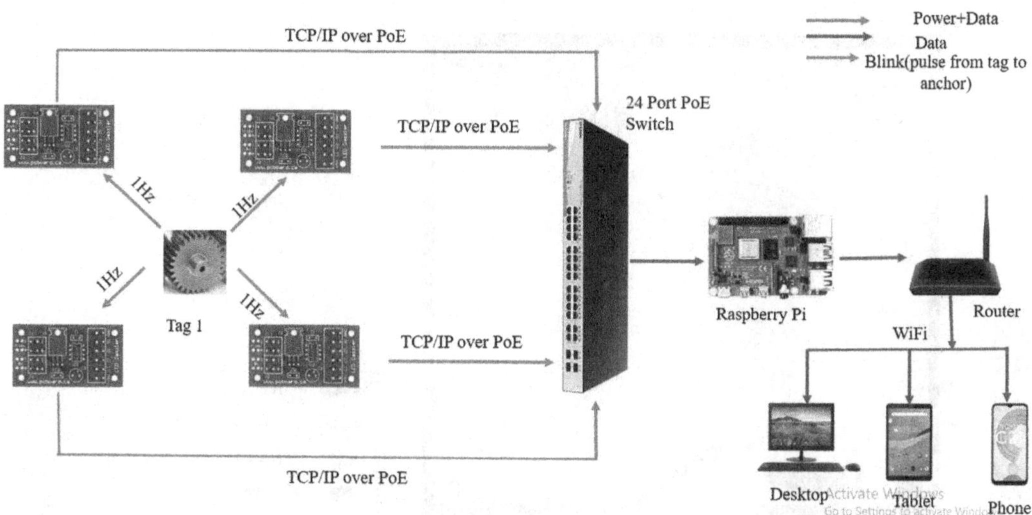

FIGURE 10.1

Simplified architecture for industry work floor.

10.3.1.2 Test bed setup

The proposed work is planned to be implemented in two areas – industry and room. The test bed is conducted in a room with a dimension of 12.5 m × 16.5 m (width × length). The industry setup will be explained in future works. The test bed setup has a total of four anchors and two tags, keeping the small dimension of the room in mind. A performance analysis would be done in the case of a test bed for improvisation in the industry setup. Four anchors are placed in each corner of the room, which provides ample coverage to the entire room. The objects placed in the room create a line-of-sight scenario. As soon as the tags are placed in the test bed setup, their information is passed to the anchors and processed as explained in the architecture. The setup is well connected over Ethernet cables. The implementation of the proposed system in an industrial arena brings about many challenges compared to a test bed setup. Since the area of implementation is small, fewer anchors and tags are required. The environment inside the industry is expected to be harsh and filled with obstacles. The results obtained in the test bed will be further analyzed and implemented considering the environment and floor plan of the industry (Fig. 10.2).

10.3.1.3 Triangulation technique

The determination of tag coordinates or assets using geometric knowledge is implemented using triangulation. Triangulation is one of the commonly used methods of position estimation, among other methods such as trilateration and multilateration. All mentioned approaches have their own method of calculating position accurately, which entirely depends on the use case of the user. Since the goal of the proposed work is to achieve higher accuracy and less complexity,

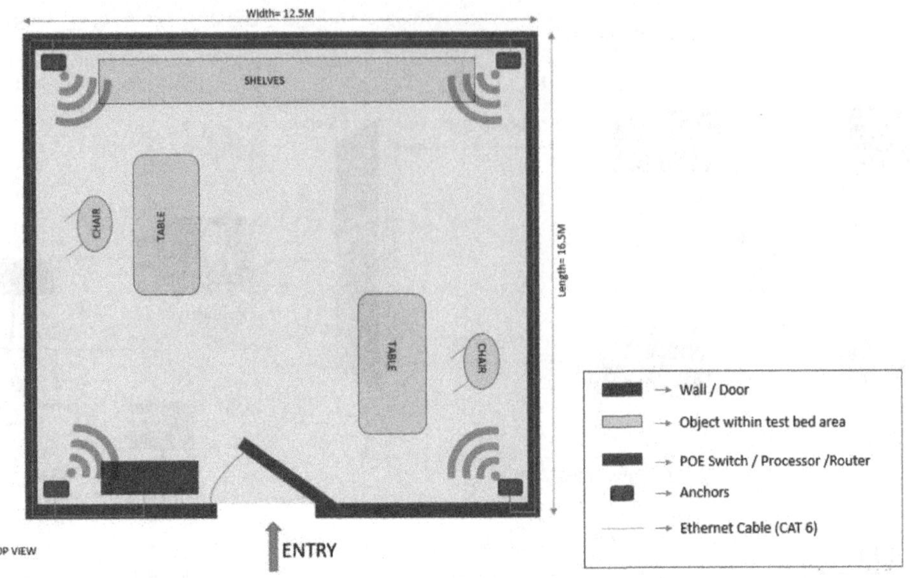

FIGURE 10.2

Test bed set up.

the triangulation algorithm has been chosen as the method of position estimation. When distance is used as a measure, it is termed time-based triangulation method. Other methods, such as angle-based triangulation and received-signal-base triangulation, also exist. The two types of time-based triangulation methods are ToA (time of arrival) and TDoA (time difference of arrival).

For experimental analysis, up to three anchors are considered with x and y coordinates for each. The object or tag coordinates are also depicted. The anchors k, l, and m when placed, collide with each other's coverage area. The method of triangulation is explained and the representation is done in 2D (Fig. 10.3).

The general Euclidean distance (r_k, r_l, r_m) can be calculated as:

$$r_k = \sqrt{[x_k - x_o]^2 + [y_k - y_o]^2}$$

$$r_l = \sqrt{[x_l - x_o]^2 + [y_l - y_o]^2}$$

$$r_m = \sqrt{[x_m - x_o]^2 + [y_m - y_o]^2}$$

Then,

$$r_k^2 - r_l^2 = -2x_k x_O - 2y_k y_o - x_l^2 - y_l^2 + 2x_l x_O + 2y_l y_O + x_k^2 + y_k^2 \tag{10.1}$$

$$r_k^2 - r_m^2 = -2x_k x_O - 2y_k y_o - x_m^2 - y_m^2 + 2x_m x_O + 2y_m y_O + x_k^2 + y_k^2 \tag{10.2}$$

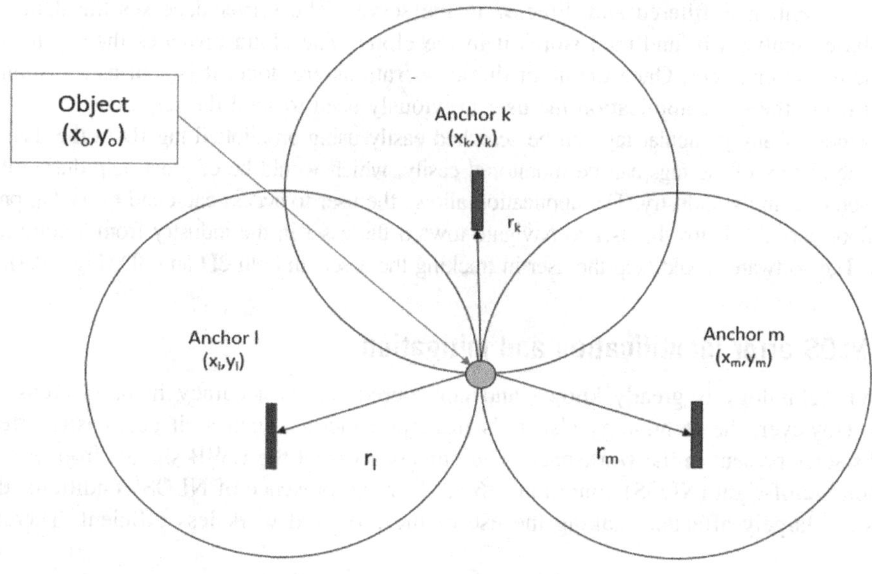

FIGURE 10.3

Triangulation technique.

On combining (10.1) and (10.2) and by extending for N pieces of anchor targets, the value of $\bar{i}$ and $\bar{j}$ can be concluded as:

$$\bar{k} = \begin{bmatrix} 2(x_2 - x_1) & 2(y_2 - y_1) \\ 2(x_3 - x_1) & 2(y_3 - y_1) \\ 2(x_N - x_1) & 2(y_N - y_1) \end{bmatrix}$$

$$\bar{l} = \begin{bmatrix} r_1^2 - r_2^2 + (x_2^2 + y_2^2 - x_1^2 - y_1^2) \\ r_1^2 - r_3^2 + (x_3^2 + y_3^2 - x_1^2 - y_1^2) \\ r_1^2 - r_N^2 + (x_N^2 + x_N^2 - x_1^2 - y_1^2) \end{bmatrix}$$

Hence the coordinates of the object in 2D can be solved by the following equation,

$$\begin{bmatrix} x_0 \\ y_0 \end{bmatrix} = \left((\bar{k})^T \bar{k} \right)^{-1} * \left((\bar{k})^T \bar{l} \right)$$

To find out the 3D coordinates a minimum of four anchors are required with x-, y-, and z-axis. In that case the object coordinates would be x_o, y_o, and z_o.

10.3.2 Asset monitoring

The anchors carry information about the pulse timings of the tag using the TDoA method. The information has been sent via the server to the local database and stored for performing operations. The user can request to access the tag information either using a web or a mobile application. Once the request is sent, it is filtered and directed to the server. The server accesses the data from the local database, analyzes it, and then stores it in the cloud. The cloud provides the tag information at any time to the end user. Once the algorithmic operations are done, it is sent as a response back to the user using the same application the user previously used to send the request.

The position of any particular tag can be searched easily using an allotted tag ID in the data analysis software. The history of the tags can be monitored easily, which would be of great help during the entire production process in the industry. The application allows the user to access each and every tag present on the work floor and also helps the user to navigate toward the asset in the industry from a common reference point. The software would help the user in tracking the assets in both 2D and 3D (Fig. 10.4).

10.3.3 NLOS error identification and mitigation

UWB band technology is greatly known and considered for its accuracy in large areas, such as industries. However, the technology also fails in certain factors, such as it gets easily affected by several obstacles present in the workspace. The objects obstruct the UWB signals from passing and set up a nonline-of-sight (NLOS) condition [36–39]. In the presence of NLOS conditions, the measurements are largely affected, making the use of the proposed work less efficient. Therefore the

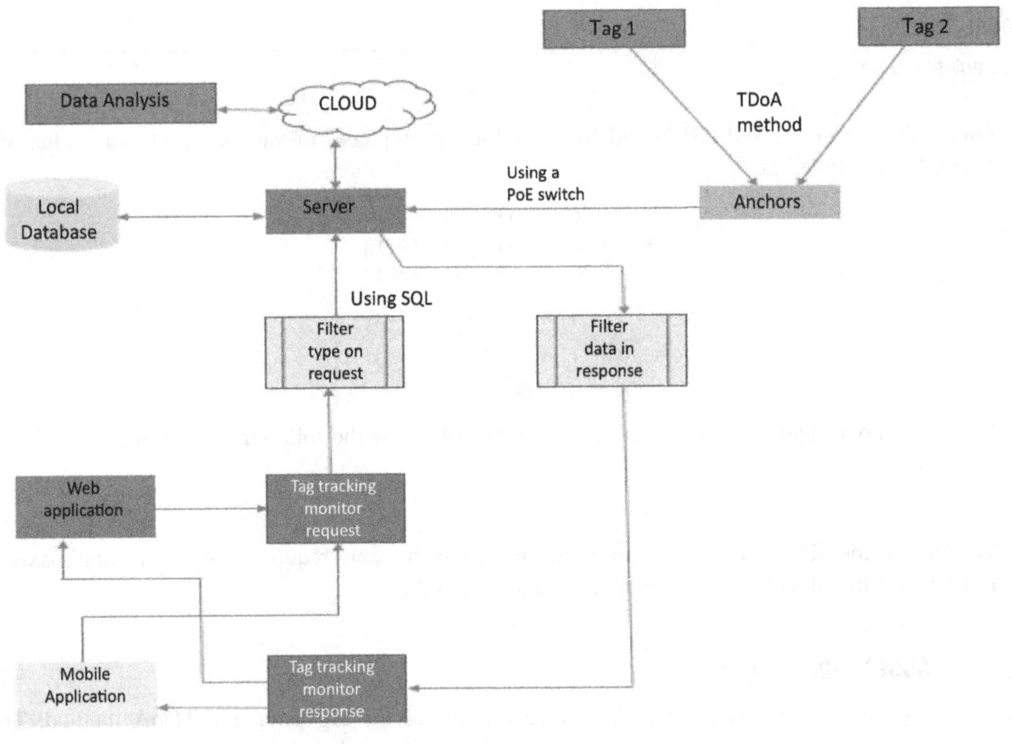

FIGURE 10.4

Simplified software architecture for the industry work floor.

authors have incorporated IMU (inertial measurement unit) with NLOS to avoid such errors. In case of scenarios where many obstacles are present, the removal of such errors is highly mandatory. The data obtained from IMU and UWB is put together to estimate the error, and an appropriate filter algorithm is used. Due to the presence of NLOS, the tag information would be scattered at various points. The measurements are made sure to fall on a definite scope. On knowing the coordinates of each tag, a filtering algorithm is advised to be used. Hence the proposed work makes use of the Kalman filter as the filtering algorithm [40,41].

10.4 Device characteristics

10.4.1 Anchors

The wall-fixed anchors play an inevitable role in the positioning process. They are static devices capable of receiving pulses from the UWB tags and forwarding them to the server for initiating the position calculation of the intended tag with the help of a master anchor. The anchors with optimal quality parameters of UWB links and strategic deployments are selected as the master anchor. A set of proposed anchors create the location infrastructure in the facility. The placement of the anchors depends on how large the industry work floor is. About 20 anchors have been placed, each being 20 m apart from each other with a range of 15–50 m. The number of anchors can be increased in case of an expansion in area localization, making the overall system fully scalable. The anchors are equipped with UWB radio, ARM controller, Ethernet interface, mini-USB, and status LED for data transmission and configuration with a dimension of $75 \times 70 \times 30$ mm. Anchors function without a battery and are powered using a PoE cable, which eliminates the need for separate power cables. In case of a power failure the USB acts as a power backup to the anchors. They are designed keeping in mind the harsh industrial environments and are resistant to dust, making them highly reliable. The status LED in anchors helps in easy identification and maintenance. The cost parameters are designed to be less, which contributes to the overall cost reduction of the end product, making it an optimal choice for low-cost industrial purposes. In Fig. 10.5 the anchor design is explained to depict a clear idea of the design involved.

10.4.2 Tags

Tags are devices attached to static and dynamic assets meant to be tracked. In this project, tags are associated with the static assets in the industry, which are meant to be tracked and monitored. The location estimation of a particular asset is done using the pulse sent out by such tags to the anchors. Tags can be designed keeping the specific purpose in mind. Personal tags act as a wireless locator for detecting and reporting real-time location. It offers a minimalistic design and can be used as a wristband, conference badge, or in clip version, whereas a vehicle tag is an upgraded tag detecting micro motion and having a prolonged battery life. The proposed work uses an industrial tag, shown in Fig. 10.6, which has high resistance and would easily adapt to harsh industrial environments. The tag is powered by a 600 mAh coin cell battery, which makes its battery life to about 2–3 years. The radio range is about 15–50 m with a dimension of about $40 \times 30 \times 15$ mm. The tags are compliant with UWB PHY 1EEE 802.15.a and are equipped with UWB chips, Li-Po/Li-Ion battery chargers, pressure sensors, and micro-USB connectors for charging. The accelerometer helps the

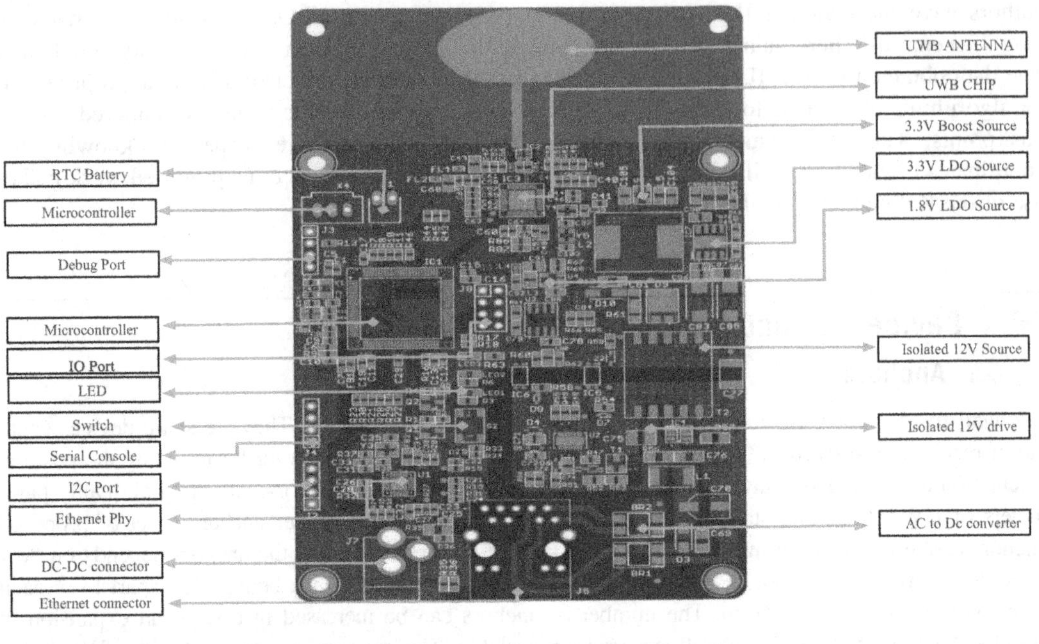

FIGURE 10.5

Anchor design in the proposed work in detail with components.

microcontroller to wake up from Deep Sleep mode when the tag is moved, which helps in prolonging the battery life (Table 10.2).

10.4.3 Decawave DW1000 UWB chip

Decawave DW1000 UWB chip, considered the world's first single-chip transceiver based on UWB technology, is the most commonly used chip for UWB-based indoor positioning systems. The module diagram is illustrated in detail in Fig. 10.7. The particular model has been the first among a series of Decawave UWB chips, which are low-priced, which is an important feature in helping to make the entire system cost-effective. The communication among the various modules of the IC is depicted in Fig. 10.7. It is based on IEEE 802.15.4 and is much sought after due to its 6.8 Mbps communication capability and a communication range of about 300 m. Capable of supporting 6 RF bands from 3.5 to 6 GHz, its communications range extends up to 290 m @ 110 kbps. It is interfaced to the microcontroller using SPI communication. The chip has the advantage of a high multipath fading immunity and supports high tag densities in real-time applications. It consumes less power and does not require the battery to be replaced for a longer period of time. The average power consumption is less than other Decawave products, including DWM 1001. However, energy consumption is higher in this product.

Decawave makes use of six radio channels in its high and low bands. The presence of such radio channels enables it to switch to any one channel in any case of interference, making it highly

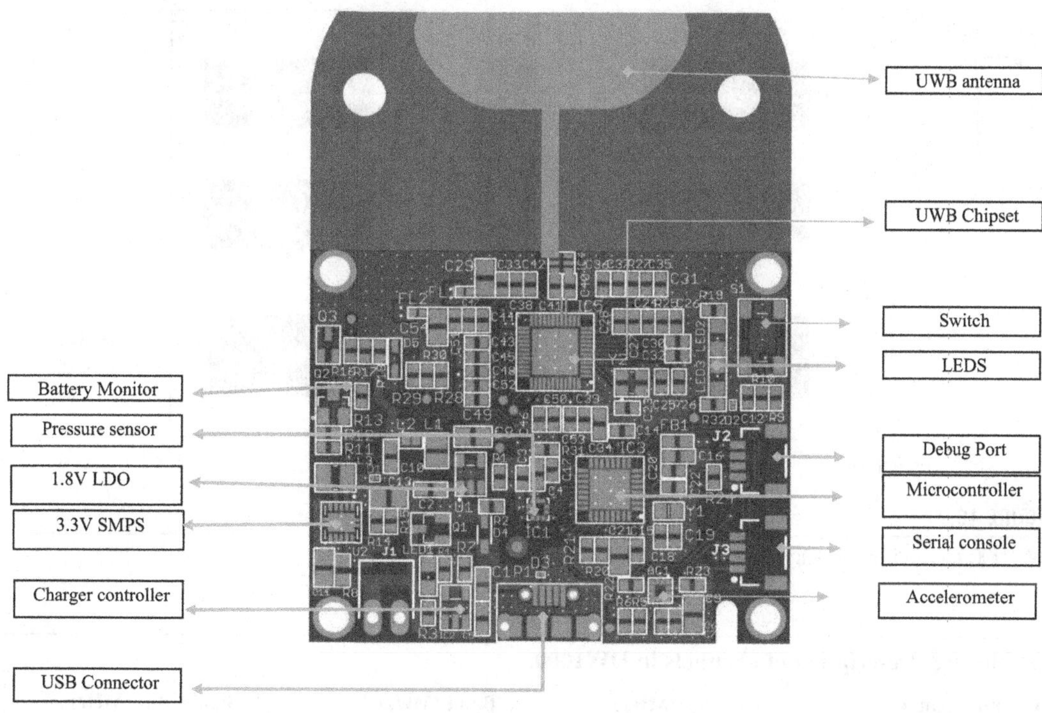

FIGURE 10.6

Tag design in the proposed work with components in detail.

Table 10.2 Comparison of various parameters in tags and anchors used in the system.

Parameters	Tags	Anchors
1. PCB material	Four-layer board epoxy resin	Four-layer board epoxy resin
2. PCB size	45 × 60 mm	74 × 130 mm
3. Main controller	STM32L041K6U6D	STM32F407VET6
4. Communication range	15–50 m (Ideal 30 m)	15–50 m (Ideal 30 m)
5. Powered by	Li-Po 3.7 V Battery	Power over Ethernet
6. Presence of sensors	Nil	Accelerometer, Pressure sensor
7. Battery life	2–3 years	As long as connected to PoE
8. Type of motion	Static	Static
9. Role in proposed work	Sends out pulses to anchors	Receives pulses and sends information to server

flexible. DW1000 makes use of the first three data rates from the IEEE 802.15.4 standard, namely 110 kbps, 850 kbps, and 6.8 Mbps. For localization purposes, channels 1, 2, 3, and 5 are used. Table 10.3 [23] (referred from one of the UWB technology Sewio websites) depicts the frequency, band, and bandwidth of each channel.

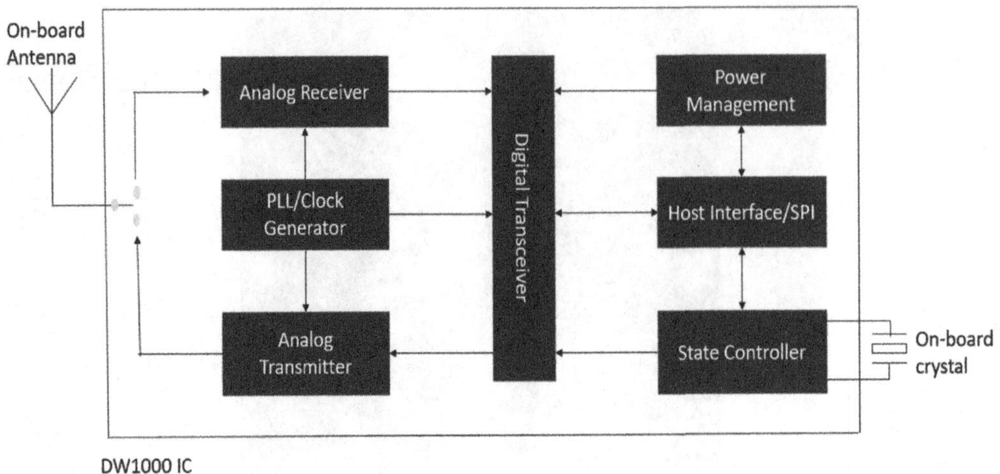

FIGURE 10.7

DW1000 module block diagram.

Table 10.3 Description of channels in DW1000.

Channel center	Frequency (MHz)	Band (MHz)	Bandwidth (MHz)
1	3495.4	3244.8–3744	499.2
2	3993.6	3774–4243.2	499.2
3	4492.8	4243.2–4742.4	499.2
4	3993.6	3328–4659.2	1331.2 (real approx. 900)
5	6489.6	6240–6739.2	499.2
6	6489.6	5980.3–6998.9	1081.6 (real approx. 900)

10.4.4 Microcontroller

The microcontroller is considered the heart of an embedded system. It consists of a memory, input/output pins, and peripherals. The input data is received by the input/output and communicated to the processor. The memory unit is responsible for stacking up the required data and using it when necessary. Two memory units are involved: One stores memory for a longer period of time while the other holds it for a shorter time. The tags are operated using the microcontroller STM32L041K6U6D and anchored with STM32F407VET6 (Table 10.4).

10.4.5 Raspberry Pi

Raspberry Pi [42], shown in Fig. 10.8, is a miniaturized card computer, that supports both Linux and Windows systems and runs specific applications on them. They play an important role in present-day

Table 10.4 Comparison of the different microcontrollers used in tags and anchors.

Parameters	STM32L041K6U6D (Tags)	STM32F407VET6 (Anchors)
1. Core	Arm 32-bit Cortex-M0 + 0.95 DMIPS/MHz	Arm 32-bit Cortex-M4 CPU with FPU 1.25 DMIPS/MHz (Dhrystone 2.1), and DSP instructions
2. Memory	32 KB Flash with ECC, 8 KB RAM	Up to 1 Mbyte of Flash memory, 192 + 4 Kbytes of SRAM, including 64-Kbyte of CCM (core-coupled memory)
3. Clock sources	1–25 MHz crystal oscillator, 32 kHz oscillator for RTC with calibration	4-to-26 MHz crystal oscillator, 32 kHz oscillator for RTC with calibration
4. Communication interfaces	5 × peripherals communication interface	Up to 15 communication interfaces

FIGURE 10.8

Raspberry Pi 3 Model B + .

embedded systems and IoT platforms to perform real-time tasks. It is often compared with embedded controllers as it is capable of supporting high-end levels of application development and performing complex tasks. The use of a development language is not restricted to C, and various other languages can still be used. The underlying hardware and upper application join hands to perform various rigorous IoT platforms. Several advantages are linked to using a Raspberry Pi compared with a general PC. First, it is capable of directly controlling underlying hardware. Second, compared with a PC, its size and cost are less. In this proposed work, we intend to make use of the Raspberry Pi 3 Model B + as a server. The server will be made responsible for the specific algorithmic functions in the proposed work to calculate the precise tag position. The PoE switch on receiving the data connects with the server to forward it. The server forwards the calculated data to the router to be sent to multiple devices.

Specifications of the Raspberry Pi 3 Model B +

- Power Input of 5 V/2.5 DC
- Supports Power over Ethernet
- Stores data and loads operating system using micro-SD port system.
- Gigabit Ethernet over USB 2.0
- 2.4 GHz and 5 GHz IEEE 802.11.b/g/n/ac wireless LAN, Bluetooth 4.2, BLE

- Extended 40-pin GPIO header
- Broadcom BCM2837, Cortex-A53 (ARMv8) 64-bit SoC @ 1.4 GHz

10.5 Experimental results

For a perfect analysis, about 50 ranging measurements were conducted from each tag for a number of 40 different positions within the coverage area of the anchors. The coordinates of the tags are mentioned in the table below (Table 10.5).

The anchors are placed at the height of 2 m from the floor to evaluate the performance of the mitigation operations explained previously in the paper. The NLOS mitigation has been run using MATLAB® software. The following figure portrays the projection of 40 different positions of the tags along with the objects present in the test bed (Figs. 10.9–10.12).

Table 10.5 Coordinates of the tags.

Beacon (*n*)	*x*	*y*	*z*
1	−0.275	3.10	2.25
2	−0.219	−1.24	20.20
3	6.15	−1.29	2.60
4	4.80	1.89	2.65

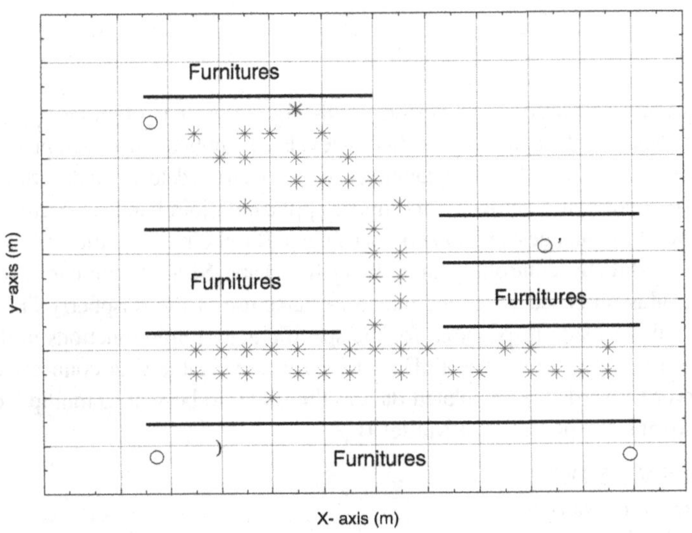

FIGURE 10.9

Projection of 40 different positions of the tags along with the objects.

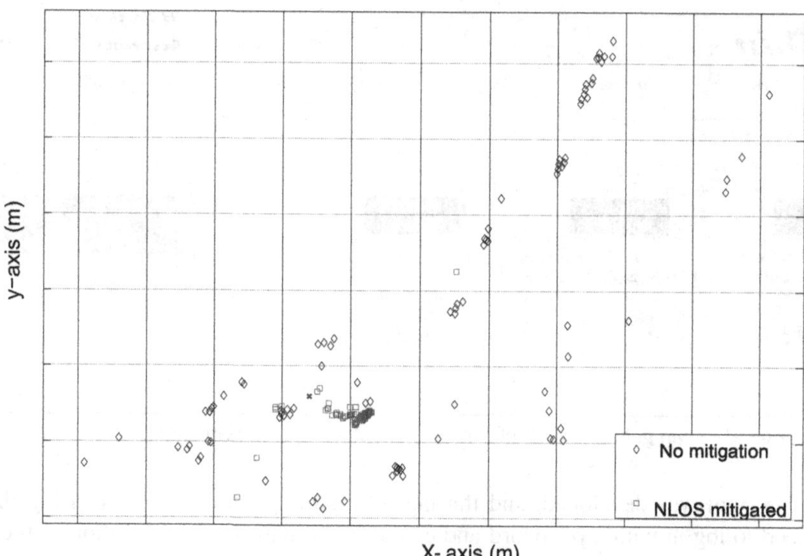

FIGURE 10.10

Positioning results for the coordinates before and after NLOS mitigation.

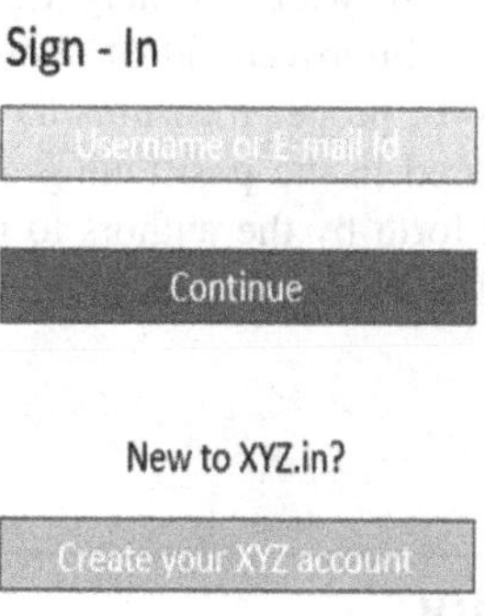

FIGURE 10.11

The user can sign in by typing a valid username and password.

The positioning results of the tags before and after NLOS mitigation is shown below. The results show that there is an improved variation of about 67% in the case of LOS (line-of-sight) and NLOS scenarios.

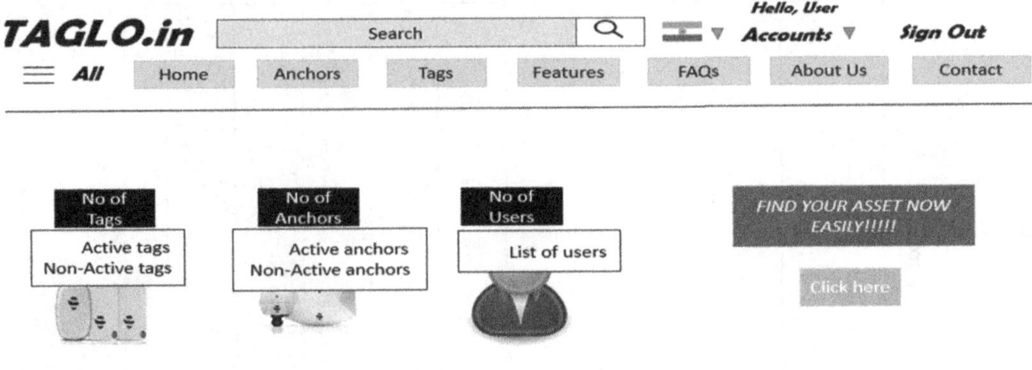

FIGURE 10.12

The user once signed in can get access to information about all tags and anchors.

The software tool was developed, and the tag information can be accessed using this software tool. Users need to log in with a password and can access the history and location of the tag.

10.6 Conclusion

Keeping in mind the importance of UWB in industries, the authors have implemented a UWB based cyber-physical system in an edge environment for indoors. The proposed work consists of a system and software architecture for precise positioning and monitoring of tags. The techniques implemented are TDoA method along with triangulation. The initial implementation took place in a room with a dimension of 12.5 m × 11.5 m (width × height). The results are shown for both LOS and NLOS scenarios. There has been an improvement of about 68% in the case of LOS scenario. Due to the limitations of the use of UWB, proper measures have been taken to mitigate them using the Kalman filter. The accuracy measured in the positioning of tags are found to be from 2 cm to 10 cm. The software tool has been put forth by the authors to indicate the tracking of assets in real time using a mobile/web application. The industry implementation will be communicated in further papers.

10.7 Characteristics summary

The features of the proposed work has been put forth together. The concluded details on the metrics, techniques, technology used, accuracy, type of assets, and applications are listed below (Table 10.6).

Table 10.6 Characteristics of the proposed system in a nutshell.

Precision obtained	From 2 cm to 10 cm
Methodology used	UWB with TDoA and Triangulation
Communication method	TCP/IP
Asset type	Static and dynamic assets
Mode of plotting	2D and 3D
Tool for monitoring assets	Via a data analysis software accessed using web/mobile application
Method of eradicating NLOS	UWB with IMU and Kalman filter
Battery	2–3 years
Consumption of power	Minimum
Places of implementation	Room with 12.5 m × 16.5 m (width × length)
Future work	Implementation in industry

Acknowledgement

The authors would like to thank DST-NRDMS funding agency for their support. (Grant no: NRDMS/UG/NetworkProject/e-13/2019 (C) P-4).

References

[1] T.A. Tran, T. Ruppert, J. Abonyi, Indoor positioning systems can revolutionise digital lean, Appl. Sci. 11 (11) (2021) 5291.

[2] Z. You, L. Feng, Integration of industry 4.0 related technologies in construction industry: a framework of cyber-physical system, IEEE Access. 8 (2020) 122908–122922. Available from: https://doi.org/10.1109/ACCESS.2020.3007206.

[3] O.C. Abikoye, et al., Application of internet of thing and cyber physical system in industry 4.0 smart manufacturing, Emergence of Cyber Physical System and IoT in Smart Automation and Robotics, Springer, Cham, 2021, pp. 203–217.

[4] G. Li, J. Xu, Z. Li, C. Chen, Z. Kan, Sensing and navigation of wearable assistance cognitive systems for the visually impaired, IEEE Trans. Cognit. Developmental Syst. (2022).

[5] M. Biswas, A. Rahman, M.S. Kaiser, S. Al Mamun, K.S. Ebne Mizan, M.S. Islam, et al., Indoor navigation support system for patients with neurodegenerative diseases, International Conference on Brain Informatics, Springer, Cham, 2021, pp. 411–422.

[6] F. Kardaş, Ö. Karal, Wireless indoor localization problem with artificial neural network, 2021 Innovations in Intelligent Systems and Applications Conference (ASYU), IEEE, 2021, pp. 1–5.

[7] J. Kunhoth, A.G. Karkar, S. Al-Maadeed, A. Al-Ali, Indoor positioning and wayfinding systems: a survey, Human-Centric Comput. Inf. Sci. 10 (1) (2020) 1–41.

[8] F. Alrefaei, A. Alzahrani, H. Song, M. Zohdy, S. Alrefaei, Cyber physical systems, a new challenge and security issue for the aviation, 2021 IEEE International IOT, Electronics and Mechatronics Conference (IEMTRONICS), IEEE, 2021, pp. 1–5.

[9] R. Verma, Smart city healthcare cyber physical system: characteristics, technologies and challenges, Wirel. personal. Commun. 122 (2) (2022) 1413–1433.
[10] S. Noel, M. Navya, D. Likitha, K. Manjula, S.K. Priya, A smart IoT based real-time system to minimize mishandled luggage at airports, 2021 5th International Conference on Computing Methodologies and Communication (ICCMC), IEEE, 2021, pp. 377–384.
[11] S. Shyam, S. Juliet, K. Ezra, Tracking and monitoring of medical equipments using UWB for smart healthcare, 2022 6th International Conference on Computing Methodologies and Communication (ICCMC), IEEE, 2022, pp. 631–637.
[12] S. Shyam, S. Juliet, K. Ezra, Indoor positioning systems: a blessing for seamless object identification, monitoring, and tracking, Front. Public Health 10 (2022).
[13] S.M. Sheikh, H.M. Asif, K. Raahemifar, F. Al-Turjman, Time difference of arrival based indoor positioning system using visible light communication, IEEE Access. 9 (2021) 52113–52124.
[14] F. Qin, T. Zuo, X. Wang, Ccpos: Wifi fingerprint indoor positioning system based on cdae-cnn, Sensors 21 (4) (2021) 1114.
[15] S. Sophia, B.M. Shankar, K. Akshya, A.C. Arunachalam, V.T.Y. Avanthika, S. Deepak, Bluetooth low energy based indoor positioning system using ESP32, 2021 Third International Conference on Inventive Research in Computing Applications (ICIRCA), IEEE, 2021, pp. 1698–1702.
[16] M. Sasikala, J. Athena, A.S. Rini, Received signal strength based indoor positioning with RFID, 2021 IEEE International Conference on RFID Technology and Applications (RFID-TA), IEEE, 2021, pp. 260–263.
[17] J. Li, G. Feng, W. Wei, C. Luo, L. Cheng, H. Wang, et al., PSOTrack: a RFID-based system for random moving objects tracking in unconstrained indoor environment, IEEE Internet Things J. 5 (6) (2018) 4632–4641.
[18] T. Obaid, H. Rashed, A. Abou-Elnour, M. Rehan, M.M. Saleh, M. Tarique, Zigbee technology and its application in wireless home automation systems: a survey, Int. J. Comput. Netw. Commun. 6 (4) (2014) 115.
[19] F.B. Islam, J.-M. Lee, D.-S. Kim, Smart factory floor safety monitoring using UWB sensor, IET Sci. Meas. Technol. 16 (7) (2022) 412–425.
[20] A. Albaidhani, A. Morell, J.L. Vicario, Ranging in UWB using commercial radio modules: experimental validation and NLOS mitigation, in: Proc. Int. Conf. Indoor Positioning Indoor Navigat. (IPIN), October 2016, pp. 1–7.
[21] F. Zafari, A. Gkelias, et al., A survey of indoor localization systems and technologies, IEEE Commun. Surv. Tut. 21 (3) (2019) 2568–2599.
[22] Q. Liu, Z. Yin, Y. Zhao, Z. Wu, M. Wu, UWB LOS/NLOS identification in multiple indoor environments using deep learning methods, Phys. Commun. 52 (2022) 101695.
[23] R. Simedroni, et al., Indoor positioning using decawave MDEK1001, in: Proc. Int. Workshop Antenna Technol. (iWAT), February 2020, pp. 1–4.
[24] E. Puschita, et al., Performance evaluation of the UWB-based CDS indoor positioning solution, in: Proc. Int. Workshop Antenna Technol. (iWAT), February 2020, pp. 1–4.
[25] S. He, X. Dong, High-accuracy localization platform using asynchronous time difference of arrival technology, IEEE Trans. Instrum. Meas. 66 (7) (2017) 1728–1742.
[26] J.P. Queralta, C.M. Almansa, F. Schiano, D. Floreano, T. Westerlund, Uwb-based system for uav localization in gnss-denied environments: characterization and dataset, 2020 IEEE/RSJ International Conference on Intelligent Robots and Systems (IROS), IEEE, 2020, pp. 4521–4528.
[27] M. Delamare, R. Boutteau, X. Savatier, N. Iriart, Static and dynamic evaluation of an UWB localization system for industrial applications, Science 2 (2) (2020) 23.

[28] P. Krapez, M. Munih, Anchor calibration for real-time-measurement localization systems, IEEE Trans. Instrum. Meas. 69 (12) (2020) 9907–9917.

[29] Apple. Apple—iPhone 11. Accessed: Mar. 31. [Online]. Available: https://www.apple.com/iphone-11.

[30] S. Shyam, S. Juliet, K. Ezra, Vehicle monitoring and localization using UWB in complex parking lot, 2022 International Conference on Sustainable Computing and Data Communication Systems (ICSCDS), IEEE, 2022, pp. 937–943.

[31] B. Venkata Krishnaveni, K. Suresh Reddy, P. Ramana Reddy, Indoor positioning and tracking by coupling IMU and UWB with the extended kalman filter, IETE J. Res. (2022) 1–10.

[32] A.L. Creţu-Sîrcu, H. Schiøler, J.P. Cederholm, I. Sîrcu, A. Schjørring, I.R. Larrad, et al., Evaluation and comparison of ultrasonic and UWB technology for indoor localization in an industrial environment, Sensors 22 (8) (2022) 2927.

[33] A.R. Jimenez, F. Seco, Comparing decawave and bespoon UWB location systems: indoor/outdoor performance analysis, in: Proc. Int. Conf. Indoor Positioning Indoor Navigat. (IPIN), October 2016, pp. 1–8.

[34] V. Djaja-Josko, J. Kolakowski, J. Modelski, TDOA estimation using a pair of synchronized DW1000 based anchor nodes, 2018 22nd International Microwave and Radar Conference (MIKON), IEEE, 2018.

[35] W. Zhao, J. Panerati, A.P. Schoellig, Learning-based bias correction for time difference of arrival ultra-wideband localization of resource-constrained mobile robots, IEEE Robot. Autom. Lett. 6 (2) (2021) 3639–3646.

[36] J. Sidorenko, V. Schatz, N. Scherer-Negenborn, M. Arens, U. Hugentobler, Error corrections for ultrawideband ranging, IEEE Trans. Instrum. Meas. 69 (11) (2020) 9037–9047.

[37] S. Djosic, et al., Multi-algorithm UWB-based localization method for mixed LOS/NLOS environments, Computer Commun. (2021).

[38] V. Rayavarapu, A. Mahapatro, NLOS identification and mitigation in UWB positioning with bagging-based ensembled classifiers, Ann. Telecommun. (2021) 1–14.

[39] K. Paszek, D. Grzechca, A. Becker, Design of the UWB positioning system simulator for LOS/NLOS environments, Sensors 21 (14) (2021) 4757.

[40] X. Yang, J. Wang, D. Song, B. Feng, H. Ye, A novel NLOS error compensation method based IMU for UWB indoor positioning system, IEEE Sens. J. 21 (9) (2021) 11203–11212.

[41] Q. Fan, B. Sun, Y. Sun, X. Zhuang, Performance enhancement of MEMS-based INS/UWB integration for indoor navigation applications, IEEE Sens. J. 17 (10) (2017) 3116–3130.

[42] Y.B. Chu, W.K. Yap, Raspberry Pi based wireless interface system for automated microfabrication in the context of industry 4.0, Advances in Electrical and Electronic Engineering and Computer Science, Springer, Singapore, 2021, pp. 117–123.

CHAPTER

Application of intelligent edge computing and machine learning algorithms in MBTI personality prediction

11

J. Anila Sharon[1], A. Hepzibah Christinal[2], D. Abraham Chandy[3] and Chandrajit Bajaj[4]

[1]Department of Computer Science and Engineering, Karunya Institute of Technology and Sciences, Coimbatore, Tamil Nadu, India [2]Department of Mathematics, Karunya Institute of Technology and Sciences, Coimbatore, Tamil Nadu, India [3]Department of Electronics and Communication Engineering, Karunya Institute of Technology and Sciences, Coimbatore, Tamil Nadu, India [4]Department of Computer Science, Center for Computational Visualization, University of Texas, Austin, TX, United States

11.1 Introduction

A cyber-physical system (CPS) is an intelligent system that employs algorithmic computing to control or monitor a mechanism [1]. CPS combines cybernetics theory, mechatronics, design, and process science in a transdisciplinary approach [2,3]. The emphasis on the embedded system, which is a popular term for process control, is more on the computational aspects than on a strong relationship between the physical and computational parts. CPS is related to the Internet of Things (IoT) in that it has a similar basic architecture. CPS has a greater coordination and integration of computational and physical elements [4]. Autonomous automobile systems, smart grids, medical monitoring, industrial systems of control, robotics systems, and avionics for automatic pilots are all examples of CPS [5]. Automotive, chemical processes, aerospace, civil infrastructure, energy, manufacturing, transportation, healthcare, entertainment, and consumer electronics all are precursors of CPS [5]. IoT devices are similar to CPS, which can help in monitoring activities like body temperature, blood pressure, heartbeat etc., through sensors. These massive data when monitored and stored can be highly useful in assisting patients with psychological treatments and mental support. The monitored data is combined with machine learning algorithms to help in predicting the personality of the patient and provide guidance accordingly.

Machine learning is a well-known technology for personality prediction widely used by researchers. Researchers can utilize machine learning to study personality traits because it provides the advantage of making predictions based on past data [6]. Businesses and organizations are increasingly investing in machine learning-based personality analysis technology. Selecting best candidates may be accomplished with the application of a machine learning algorithm, and a human error analysis process can reduce work performance in terms of personality.

Intelligent Edge Computing for Cyber Physical Applications. DOI: https://doi.org/10.1016/B978-0-323-99412-5.00003-4

Personality can have an impact on people's emotionally meaningful experiences with situations [7]. The MBTI determines personality as a set of characteristics that describes the possibility of a person's behavior, feelings, and thoughts being unique. The characteristics of a person change with time and under different situations. Personality is a collection of traits and standards that combine to create a person's nonidentical personality. The Big Five model (five-factor model) [8], Myers–Briggs Type Indicator (MBTI) [9], and Carl Jung's Theory of Personality Types [10] are models used to describe personality, of which the MBTI and Big Five are presently prominent among researchers. MBTI is a more robust model than other models since it has a wider range of applications in diverse areas, yet it has some reliability and validity concerns. We chose the MBTI personality model for this work because of its popularity and ability to be used in a variety of industries.

Psychologists and sociologists have long been interested in personality research. Psychiatrist Carl Jung's research was based on the MBTI model using psychological types in the 1920s [11]. The MBTI tool is also known as a type table, as shown in Fig. 11.1 [12].

The MBTI classifies personality into 16 types of which INTP (introverted, intuitive, thinking, and perceiving) is one type. We can categorize a person's behavior or desire based on the label, and the machine can learn more information. The 16 personality types are combined to show four dimensions of personality preferences: S - Sensors, N - Intuitives, T- Thinkers, F - Feelers, J - Judgers, P - Perceivers and the dimensions include: (E/I), (S/N), (T/F), and (J/P), as depicted in Fig. 11.2 [13]. Each dimension corresponds to two distinct personalities.

We may use Term Frequency-Inverse Document Frequency to analyze user's MBTI types based on their social media posts, such as image, video, or other links (TF-IDF). We could make use of the TF-IDF as a technique to measure and detect a person's most well-known terms. Other potential properties derived from text data, in addition to URLs, include words, actions, ellipses, number of words, emoticons, hashtags, and many others. These supplementary characteristics play an important role, which may be linked to different personality types. When social media users are classified according to one of the MBTI personality types, their linguistic contents, such as the amount of words or ellipses, provide additional personality characteristics to the person.

The Five-Factor or the Big Five Model is the most widely used academic personality model. The relationship between Big Five and MBTI personality traits are quite substantial. While multiple

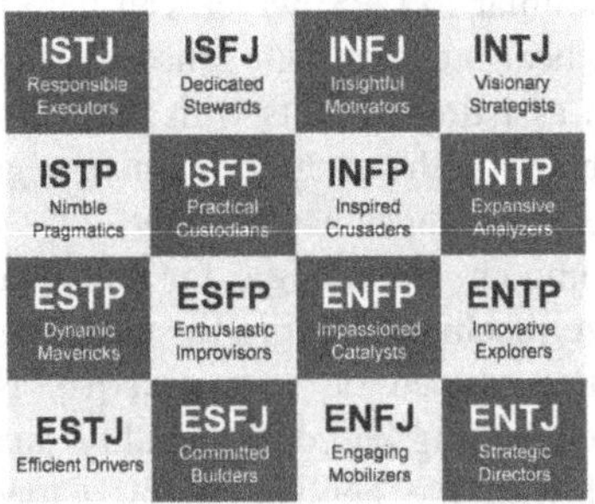

FIGURE 11.1

The Myers Briggs Type Indicator.

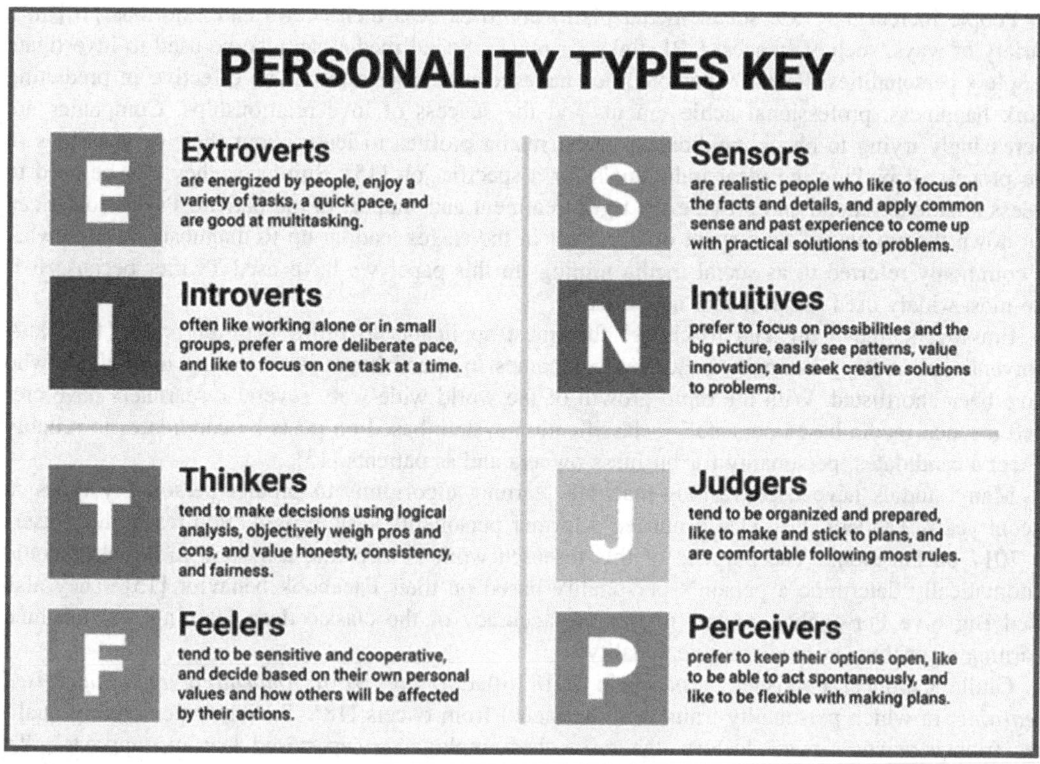

FIGURE 11.2

Key personality types.

Table 11.1 Similarities between Myers–Briggs and Big Five.

Myers–Briggs	Big Five
Extraversion (E)	Extraversion
Intuition (N)	Openness
Feeling (F)	Agreeableness
Perceiving (P)	Consciousness

studies have found a strong relationship between MBTI and Big Five, the information in this paper is mostly obtained mostly Adrian Furnham and colleagues' major study (with over 900 participants) [14]. The results of the MBTI were compared with those of the Big Five's Revised NEO-Personality Indicator in this study (NEO PI-R). The Big Five and the MBTI have a lot in common, with four of the five components exhibiting substantial connections with particular preference for MBTI as it is depicted in Table 11.1.

People increasingly use social media platforms to express their views and emotions [10] in a variety of ways, such as images, URL links, or music. Social media can also be used to investigate people's personalities. People's personalities have been demonstrated to be effective in predicting work happiness, professional achievement, and the success of love relationships. Companies are increasingly trying to check candidates' social media profiles to learn about their personalities in the process of picking the ideal individuals for a specific job [15]. Similarly, they can be used to assess a patient's mood and provide the right treatment and support to the patient. Psychologist can cut down the amount of time spent on a patient in the stages leading up to diagnosis through what is commonly referred to as social media mining. In this paper we have used Twitter because it is the most widely used social media networks.

Employers find it difficult to choose the finest applicants to fulfill their purposes [16,17]. A conventional strategy normally behooves companies to spend hours interviewing individuals who have been shortlisted. With the rapid growth of the world wide web, several researchers have created a social media-based personality classification system based on posts by candidates to reliably detect a candidates' personality for business owners and as patients [15].

Many studies have used various machine learning algorithms to predict personality types in recent years. Tandera et al. [15] produced a former personality system prediction research on users in 2017 on Facebook. The purpose of this research work is to create a system in which it could automatically determine a person's personality based on their Facebook behavior [15]. They also used Big Five Personality models to test the accuracy of the classic deep learning and machine learning algorithms in predicting personality.

Giulio Carducci published a paper in 2018 titled *Using Word Embedding and Supervised Learning*, in which personality traits were extracted from tweets [18]. To derive personality qualities from a person's tweet history, the researcher employed a supervised learning approach. To anticipate the Big Five Personality model, they built three algorithms in machine learning, such as logistic regression, LASSO, and support vector machine (SVM). In 2020 Mohammad Hossein Amirhossein published a paper titled *Personality Type Prediction Using Machine Learning in the MBTI* [19]. Based on the MBTI personality type indicator, the study established a new machine learning method for automating the identification of metaprogrammes and the predicting of personality types [19].

In this research, a method that can predict your personality based on Twitter data is proposed to estimate the personality of applicants. The extracted twitter comments of each person is compared according to their written posts with relevance on the most prominant keywords with various emotions. The slang of sentences and the written lines temperament are highly considered while extracting the twitter posts of a candidate.

Machine learning is used to mine the user attributes and discover patterns. In the suggested strategy, large volumes of personal behavioral data are collected. By processing numerous attributes, this system automatically analyzes candidates' personality qualities, eliminating the lengthy procedure necessary for a method that can predict your personality through the traditional approach. The rest of this paper is organized as follows: Section 11.2 discusses relevant work; Section 11.3 discusses proposed model; Section 11.4 discusses materials and procedures; Section 11.5 discusses the findings of the experiments; and Section 11.6 wraps up the study.

11.2 Literature survey

In the field of personality analysis, researchers have used a variety of machine learning techniques. Almost all research in this subject involves numerous processes, incorporating data collection, preprocessing, feature extraction, and classification to establish the model's accuracy. The MBTI and associated works from prior researchers employing algorithms of machine learning are highlighted in the following section.

11.2.1 Machine learning

Machine learning is a subclass of artificial intelligence that enables frameworks to learn from each other to take in and improve for a fact without being expressly stated changed [20]. Reinforcement learning, unsupervised learning and supervised learning are the three machine learning algorithms used in this paper. Supervised and unsupervised learning are the two most common and widely used strategies. Supervised and unsupervised learning are the two most common and widely used strategies. A supervised learning algorithm is made up of two variables: training data(input data) and the outcome(target) variable. The prediction model's desired output is determined by a set of features in the input [21]. Logistic Regression, Decision Tree and Linear Regression are some examples of supervised learning algorithms. Classification is used when the output variable is organized into categories, the machine learning is used to make a prediction about the outcome of the given sample. Support Vector Machine (SVM), Naive Bayes Classifier, and K-Nearest Neighbor (KNN) are examples of classification algorithms.

Unsupervised learning algorithm is a population-gathering technique where examining unlabeled data, can reveal hidden structures. Mean Shift, K-Means, and K models are examples of such algorithms. Reinforcement learning is a set of algorithms that use a trial-and-error strategy to continuously train data in order to make certain decisions. This form of learning can be used in various situations including searching by trial and error with a delayed reward [21]. This method will attempt to comprehend the greatest possible information by analyzing previously trained sample data in order to make the best judgement. Markov Decision Process and Q Learning are two examples of reinforcement learning algorithms.

For years, Facebook have used system for Predicting Personality, which can automatically determine the user personality based on their activities in Facebook [15]. The Big Five Personality model is used by Facebook to properly estimate a one's personality based on their personality attributes. Conscientiousness, Extraversion, agreeableness, openness and neuroticism are among the traits that can be uncovered using this model. The researchers employed two datasets to predict the personality of the consumers in this investigation. The first dataset contains my Personality project samples, whereas the second dataset contains data that was personally developed. Before moving on to the next stage, the documents produced in the English language are rectified in the preprocessing stage. Symbols, stemming, spaces, lowering case, names, URLs, and deleting stop words are all steps of the preprocessing process. Slang terms and unconventional words are manually substituted in a separate preprocessing stage for data in Bahasa Melayu before the texts are translated to English.

To attain optimum accuracy in this classification procedure, multiple sets of experiments were carried out, utilizing algorithms for deep learning and conventional machine learning technique for forecasting the personality type of candidates for a certain employment role. Gradient Boosting, Linear Discriminant Analysis, Logistic Regression, Naive Bayes and Support Vector Machine (SVM)are examples of traditional machine learning techniques (LDA). Meanwhile, four architectures were employed in deep learning implementations: Gated Recurrent Unit (GRU), Long Short-Term Memory (LSTM), Multi-Layer Perceptron (MLP), and 1-Dimensional Convolutional Neural Network (CNN 1D). Experiments on classic machine learning techniques revealed that average accuracy of the LDA algorithm is the highest in myPersonality dataset. Aside from that, with a manually acquired dataset, The average accuracy of the SVM algorithm is the greatest (Despite the fact that the difference between different algorithms isn't considerable). Meanwhile, experiments and tests on deep learning algorithms concluded that the MLP design has the highest average accuracy in the myPersonality dataset, while the LSTM + CNN 1D structures had the highest accuracy in manually obtained dataset. Finally, deep learning techniques can be used to improve the accuracy of datasets, even for qualities with low accuracy. This is because very small number of datasets are used in this investigation.

11.2.1.1 Personality traits assessment through tweets using word embedding and supervised learning

Twitter is used as a source for determining personality traits in our work. In this is widely used platform for social media, people is using to express diverse elements of life, including personality, is a rich textual data source and user behavior. People freely express their emotions, moods, and viewpoints, resulting in a large and useful personal data collecting that may be used for diverse of applications [18]. Aside from that, there was a latest study that created a questionnaire for personality traits called the Big Five Inventory (BFI) personality test. It contains 44 brief statements with the Likert scale of five-level that may properly evaluate the five personality traits as well as their six underlying components. After that, the Twitter handles of 26 panelists were requested as well as was asked of to complete a questionnaire. Mention removal, hashtag removal and URL removal are all textual characteristics of the user that are removed during the preprocessing stage. Aside from that, retweets with no new content were also removed. Then each tweet vector was fed into the trained model separately to get a prediction, and then averaged the results to get the end personality trait score. Researchers investigated various machine learning algorithms and their results in order to find the highest performing predictive model. The mean squared error is used as the loss function for machine learning algorithms that are assessed using the training set. The learning model (SVM) was also compared to two standard algorithms (LASSO and Linear Regression) which are the state-of-the-art approaches to personality prediction. The results indicated that the SVM classifier could accurately classify the personality of Twitter users while also achieving a decreased in mean squared error. LASSO and Linear Regression models within the myPersonality Gold Standard data, low discriminative power tends to forecast personality characteristics that are near to the average.

11.2.1.2 Personality type prediction based on MBTI by machine learning approach

The MBTI is a four-dimensional personality assessment that incorporates 16 distinct personality types. These fundamental dimensions define a person's preferences [19]. E/I, S/N, T/F, and J/P are the four dimensions that are also known as basic meta-programmers. For each dimension, there are two types of personalities. Based on the MBTI [19], this work predicted a person's personality type. Data are data from a discussion board on the internet during the preprocessing stage and used NLTK to classify the MBTI types. Then, we text lemmatized, which is the act of converting twisted or changed form of words into their base words where divided 16 personality types are divided into four categories (dimensions)., In accordance with the MBTI personality model, each of these binary classifications represents a different component of a personality.

The Gradient Boosting Model was constructed after the preprocessing stage. At this point, The each individual MBTI type indications were trained., and the data was divided into training and testing datasets. Training data was used to fit the model, while testing data was utilized to make predictions. They then employed a recurrent neural network, which is an existing technology, to calculate the accuracy of the prediction. According to the results, the XGBoost classifier, which is based on Gradient Boosting, outperformed the recurrent neural network.

To ensure that the recent strategy is superior and developed correctly, we must evaluate existing systems that have been put in place before we build it. We compared the systems using the personality model and approach that were already in place. A comparison of available techniques is shown in Table 11.2.

We had developed the new approach based on an analysis of current systems to produce better personality results and more reliable data. Additionally, increasing the dataset size may result in a more accurate forecasting. We used Twitter as a social media platform. Where tweets are collected from the Kaggle repository as our dataset for this study.

11.2.1.3 Stochastic gradient descent

Stochastic Gradient Descent (SGD) [22] is a quick and easy way to fit linear classifiers and regressors to convex loss functions, similar to (linear) Support Vector Machines and Logistic Regression [23]. SGD has been used to solve large-scale, sparse machine learning issues that are common in text categorization and natural language processing [24]. The SGD Classifier's advantages are Easy to implement and have High Efficiency [25]. The disadvantages are: it needs various hyperparameters like the number of iterations and regularization parameter and SGD is highly sensitive to feature scaling [26].

Table 11.2 Previous machine learning-based personality prediction research.

Studies	Personality model	Method
Tandera et al. [15]	Big Five Personality Model	Deep Learning, Traditional machine learning
Carducci et al. [18]	Big Five Personality Model	SVM Classifier, Linear Regression, LASSO
Amirhosseini and Kazemian [19]	Myers–Briggs Type Indicator	NLTK, XGBoost
Yoong et al. [9]	Myers–Briggs Type Indicator	Decision Tree
Pratama and Sarno	Myers–Briggs Type Indicator	Naive Bayes, KNN and SVM

11.2.2 Cyber-physical systems: Internet of Things

Cyber-physical systems (CPS) has revolutionized individualized health care, traffic flow management, emergency response, delivery, electric power generation, and a slew of other fields that are only beginning to be imagined. CPS are made up of interconnected digital, analog, physical, and human components that are designed to work together using integrated physics and logic [27]. Other terms you could hear while talking about CPS technology such as these and others are: Internet of Things (IoT), Smart Grid, Smart Cities, Industrial Internet and "Smart" (e.g., Homes, Buildings, Hospitals, Cars, Manufacturing, Appliances).

IOT application used here is the mobile phone which collects the massive useful datas from the user. Smartphones, on the other hand play a significant part because many IoT devices can be managed, via a smartphone app. IoT devices have sensors and mini-computer processors that use machine learning to act on the data acquired by the sensors. Essentially, IoT devices are mini computers, connected to the internet. Machine learning is the process by which computers learn in the same manner that humans do by collecting data from their environment and it is what allows IoT devices to become smart. This data can help the machine learn your preferences and adjust itself accordingly [28].

11.2.2.1 Support vector machine

Support-vector networks [1] is a machine learning technique that uses supervised learning models to examine the given data for categorization and regression analysis [29]. A support-vector machine, in more technical terms, is a infinite or high-dimensional environment, forms a hyperplane or series of hyperplanes that can be used for regression, classification, or other tasks such as outlier detection [30].

Intuitionally, the hyperplane with the largest distance between the nearest training data point and the any of the classes functional margin attains has the decent separation, because larger the margin, lower the classifier's generalization error [31]. Even if the initial problem is expressed in a finite-dimensional space, the sets to distinguish are frequently not linearly able to be separated in that space. As a result, it was suggested that the actual finite-dimensional space be transferred into a much higher-dimensional region, purportedly making the segregation easier [32]. To keep the computational load less, SVM systems uses mappings that ensure the dot products of two input data vectors which in turn can be, defining them in terms of a kernel function. They can be simply computed in terms of the original space's variables that has been chosen for the problem [33]. In higher-dimensional space, hyperplanes are defined as a set of points in a space whose dot product with a vector is constant, and such vectors of sets are said to be orthogonal and minimal.

SVM can be used with various kernel function of linear as well as multi-dimensional function. It has been proved to be the state of art method for classification and regression problems. It is a technique which can be used both in supervised and unsupervised technique.

11.3 Model architecture

The proposed system contains two phases: At first phase, the classification phase where the personality of the individual is identified through tweet feeds and using SGD algorithm and second phase where the individual's actions are monitored using body sensors like thermal sensors, measuring

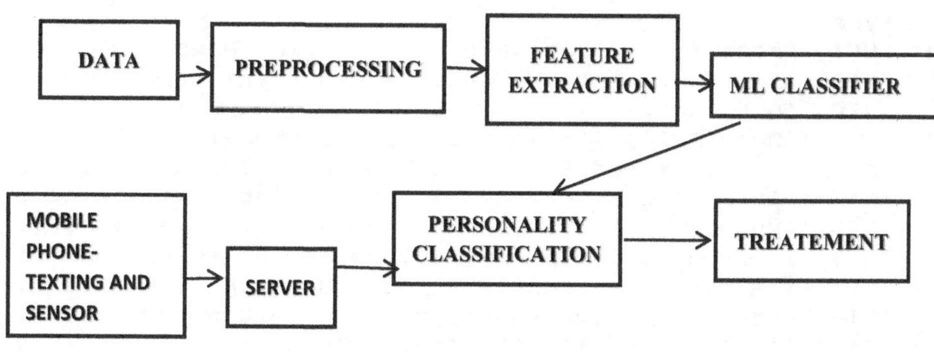

FIGURE 11.3

Architecture diagram of the proposed model.

blood pressure and heartbeat which are monitored by the mobile phone in their actions and texting process. In the second phase the various body levels are monitored and their texting pattern are t collected and send to the server for matching the current phenomenon with the action. Then the server uses a machine learning algorithms to classify the emotions generated which in turn compared with the personality of the person and proper treatment is immediately recommended. The Fig. 11.3 mentioned below shows the architecture. The whole process is stored in cloud for further proper diagnosis by the psychologists in the future checkup.

11.4 Materials and methods

The method to build our proposed architecture for personality classification using machine learning and the assistance to support through mobile phones and sensors to candidates/patients is discussed in this part. This methodology aids in the organization of the psychological patient assistance-modal development process.

11.4.1 Model development

11.4.1.1 Dataset

We used the Kaggle repository [34] to acquire the data. As shown in Fig. 11.4, the dataset in this work has 8675 rows of data in two columns.

The data in each row contained the following information about an individual:

- Type: The four-letter MBTI code/type for each individual
- Posts: Each of the person's latest 50 Twitter posts. "|||" (3 pipe character) separates each entry.)

Another dataset in Ref. [35] provided us with that of 2017. The data gathering was done in two stages. In the first step, users completed a series of questionnaires that classified them according to their MBTI personality type. They where allowed to chat publicly with other forum users in the second phase. Thus, vast number of various diverse personality type data based on MBTI type was created as a

```
  type                                                  posts
0 INFJ  'http://www.youtube.com/watch?v=qsXHcwe3krw|||...
1 ENTP  'I'm finding the lack of me in these posts ver...
2 INTP  'Good one  _____   https://www.youtube.com/wat...
3 INTJ  'Dear INTP,   I enjoyed our conversation the o...
4 ENTJ  'You're fired.|||That's another silly misconce...
5 INTJ  '18/37 @.@|||Science  is not perfect. No scien...
6 INFJ  'No, I can't draw on my own nails (haha). Thos...
7 INTJ  'I tend to build up a collection of things on ...
8 INFJ  I'm not sure, that's a good question. The dist...
9 INTP  'https://www.youtube.com/watch?v=w8-egj0y8Qs||...
*****************************************
<class 'pandas.core.frame.DataFrame'>
RangeIndex: 8675 entries, 0 to 8674
Data columns (total 2 columns):
type     8675 non-null object
posts    8675 non-null object
dtypes: object(2)
memory usage: 135.7+ KB
None
```

FIGURE 11.4

MBTI personality type dataset.

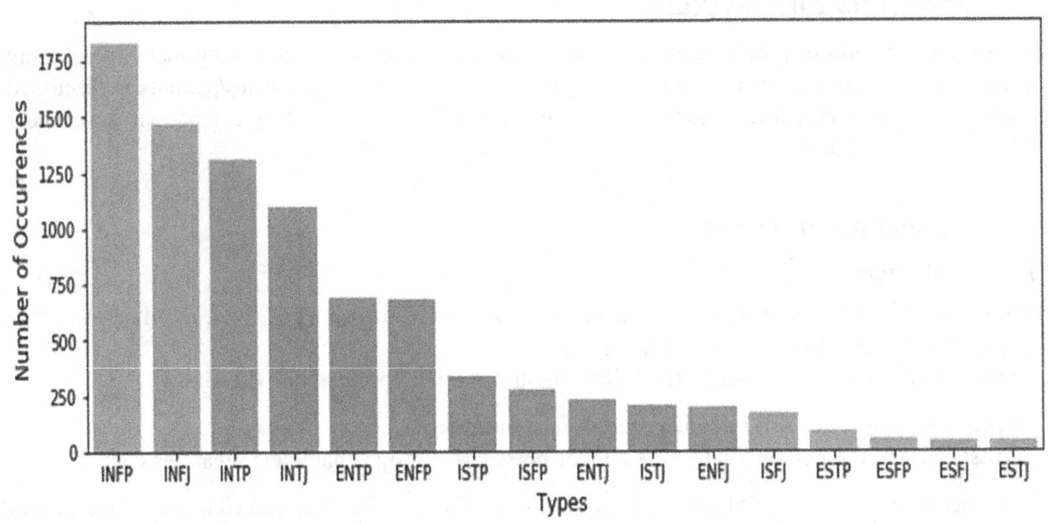

FIGURE 11.5

The number of occurrences of each MBTI personality type in the dataset.

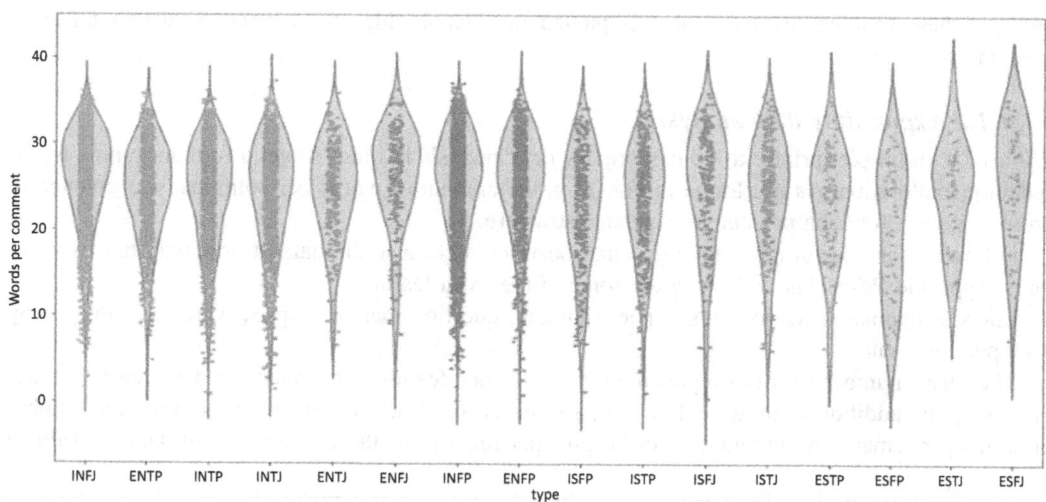

FIGURE 11.6

For each personality type, the number of words per comment is calculated.

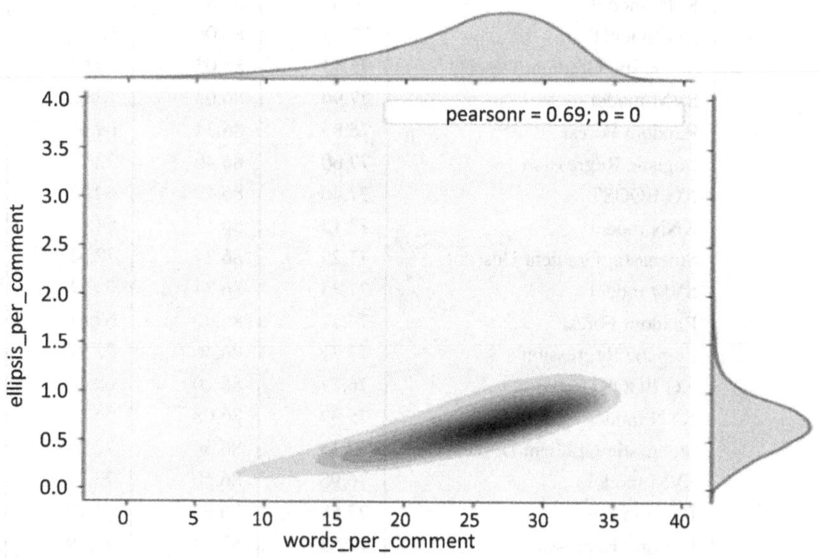

FIGURE 11.7

Pearson correlation.

result of these chatting sessions and it is plotted in a graph (Fig. 11.5) based on their number of occurances.

11.4.1.2 Exploratory data analysis

We used a violin plot printer to do exploratory data analysis in order to get a further graphic representation exploration. As illustrated in Fig. 11.6, the amount of words per comment was counted to provide a sense for each personality sentence structure.

Following that, seven new features were constructed because the dataset now only has two features, Type and Posts. The following are some of the extra features:

Links, exclamation marks, music, per comment, question marks, ellipsis, words and images by each per comment.

The mean number of words, punctuation, and other features are determined for each feature. Following the addition of these features, we looked at the Pearson association between the number of words per remark and the number of ellipses per remark for the entire group of data to evaluate

Table 11.3 Results of Models in all Training-Testing data splits ratio: 70%–30%, 80%–20%, 60%–40% and 90%–10%.

Training-testing	MODEL	I/E	N/S	F/T	J/P
70–30	Random Forest	77.68	86.03	67.87	61.33
	Logistic Regression	77.54	89.06	72.44	64.51
	KNN model	76.67	85.82	54.70	40.20
	XG BOOST	77.65	86.06	68.77	64.83
	Stochastic Gradient Descent	76.53	86.03	70.63	64.76
	SVM model	77.96	86.03	72.62	65.87
80–20	Random Forest	76.83	86.34	69.68	63.86
	Logistic Regression	77.60	86.46	73.03	65.13
	XG BOOST	77.46	86.40	69.57	64.73
	KNN model	77.12	86.17	54.47	61.38
	Stochastic Gradient Descent	77.29	86.34	72.51	65.01
	SVM model	77.35	86.34	73.31	66.74
60–40	Random Forest	76.77	86.40	67.41	61.18
	Logistic Regression	77.71	86.46	72.97	63.63
	XG BOOST	76.77	86.40	68.44	63.89
	KNN model	75.99	86.08	54.87	60.14
	Stochastic Gradient Descent	77.35	86.34	72.28	65.53
	SVM model	76.95	86.40	73.29	64.70
90–10	Random Forest	77.88	86.98	71.20	62.90
	Logistic Regression	77.53	87.21	74.19	64.63
	XG BOOST	78.80	87.10	69.59	64.29
	KNN model	78.34	86.75	55.65	39.06
	Stochastic Gradient Descent	78.23	86.98	73.27	65.21
	SVM model	78.23	86.98	74.88	66.01

how the unprocessed data looked and how the features differentiated across the four MBTI types (see Fig. 11.7). Visualization and correlation of the MBTI personality types uses the Python language data visualization library 'Seaborn,' as well as the Python 2D plotting library 'Matplotlib where used to implement.'

Pearson correlation values (pearsonr = 0.73) are provided for each personality. In Figs. 11.10 and 11.14, an ellipsis is associated with 72% of the terms. Aside from that, 64% of the terms in Figs. 11.12, 11.15, and 11.18 are connected with an ellipsis. In contrast, 74% of the terms in Figs. 11.13 and 11.16 are associated with an ellipsis.

Fig. 11.6 shows that the number of words per remark and the number of ellipses per comment are highly correlated. An ellipsis is associated with 69% of the words. It was plotted in comparison with the ellipses per comment and word per remark, a pair plot and joint plot on the association variables for the various types of personality in Table 11.3 to evaluate the personality type with highest correlation. Figs. 11.8–11.13 shows the relationship between the number of ellipsis per

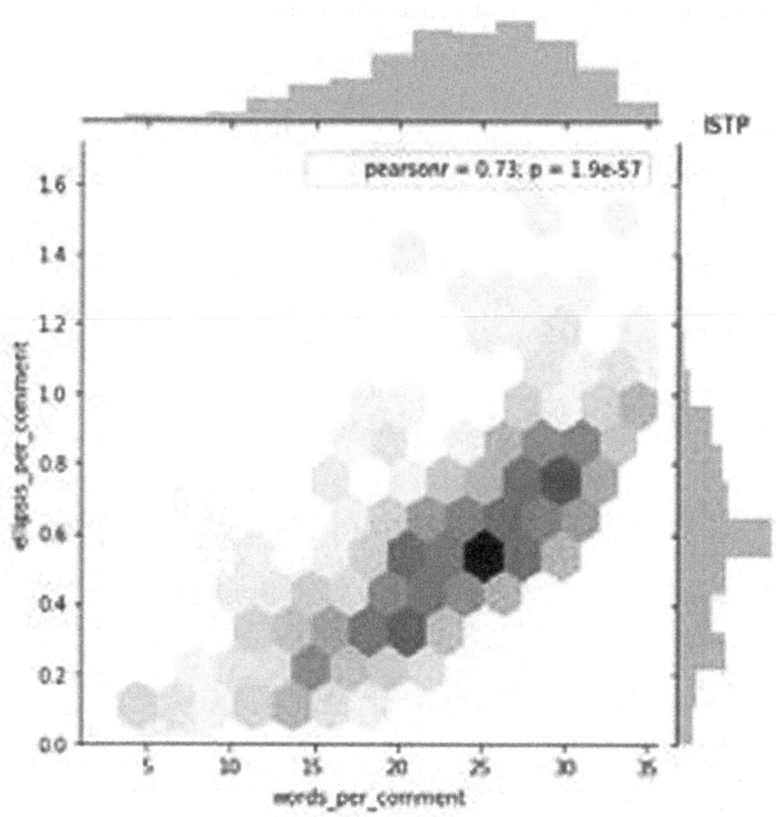

FIGURE 11.8

ISTP type.

comment and the number of words per comment for the personality types ISTJ, ISTP, ISFP, ISFJ, INTJ and INTP. Finally, The correlation between ellipsis per comment and words per comment for INFJ INFP, ENFP, ENTP, ENFJ and ENTJ personality types is depicted in Figs. 11.13–11.20.'

Table 11.3. For each MBTI personality type – Pearson Correlation For Ellipses Per Comment Vs Words Per Comment.

From Fig. 11.20, for ellipses and word per comment of each is listed. The top three largest correlation values are:

- INFJ: The advocate – Introversion Intuition Feeling Judging
- INTP: The Thinker – Introversion Intuition Thinking Perceiving
- ENFP: The Inspirer – Extroverted Intuition Feeling Perceiving

Every MBTI type have a unique relationship between ellipses and words per comment. based on the exploratory data phase. The correlation identifies the degree to which each trait is influenced by another. The strongest correlation was found among INFJ, INTP, and ENFP, which is a good sign for training data and developing machine learning models.

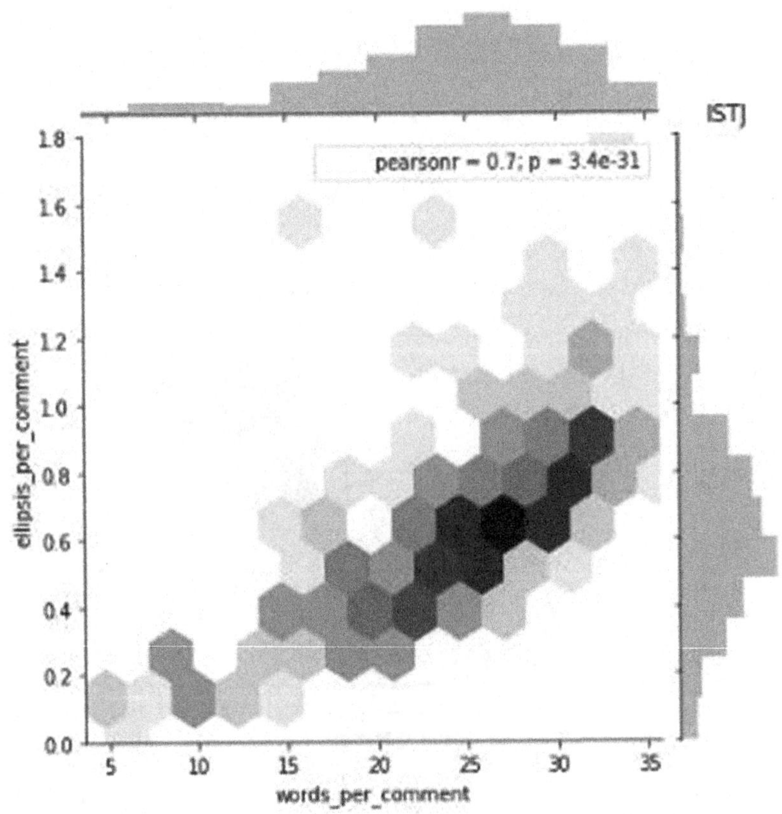

FIGURE 11.9

ISTJ type.

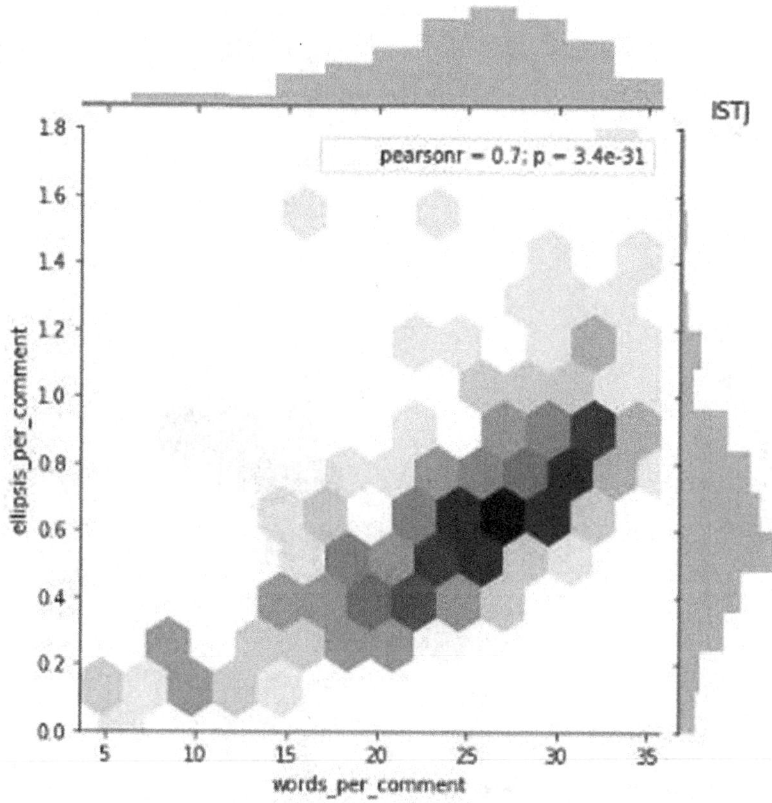

FIGURE 11.10

ISFP type.

The Pearson correlation coefficient can be utilized to assess the strength of links and variables. The value of the coefficient, which might range between −1.00 and 1.00, must be found in order to comprehend how substantial the association between two variables is. The equation of Pearson equation is given as, where, r stands for the correlation coefficient., xi stands for values of the x-variable in a sample, x bar stands for the mean of the values of the x-variable, yi stands for the values of the y-variable in a sample and y bar stands for the mean of the values of the y-variable.

The correlation coefficient between personality type identifiers is shown in Fig. 11.22. The distribution of count of words of the identifiers throughout the data is shown in the Figs. 11.21 and 11.23.

11.4.1.3 Data preprocessing

Four newly created columns where developed totally to split the respondents based on the four MBTI dimensions: E/I, S/N, T/F, and J/P to gain more insight into the dataset. The goal of the procedure is to increase the accuracy of the results.

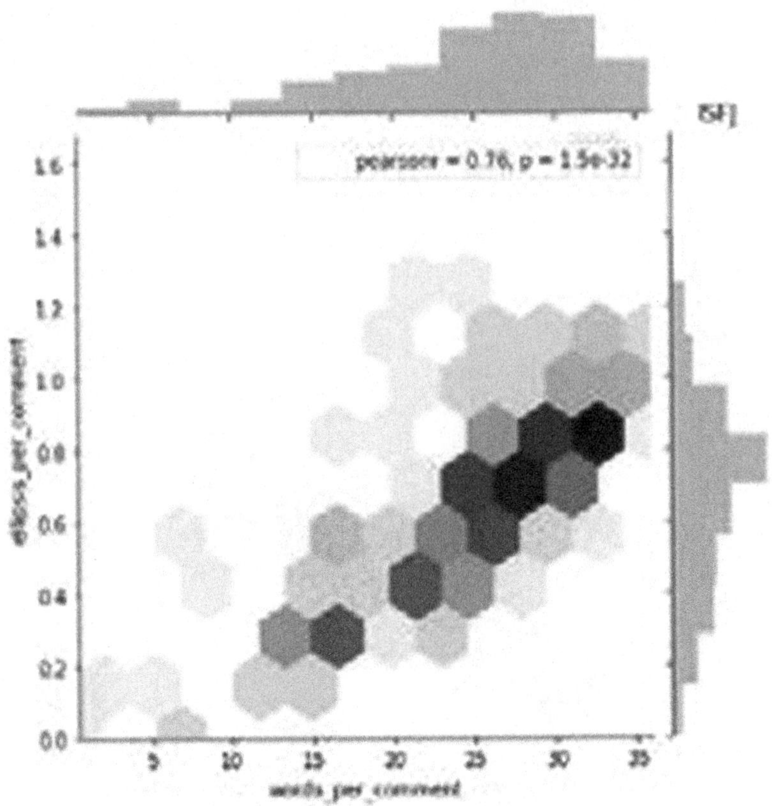

FIGURE 11.11

ISFJ type.

In this preprocessing step, we used the word2vec approach. Word2vec is a word embedding algorithm that generates vector representations of words. The word corpus is "gensim" imported from word2vec library. The model implemented of word2vec is CBOW(Continuous Bag of Words). CBOW is the neural network examines the surrounding words and predicts the word that appears in the middle. The batch size used is 512 throughout the implementation. We turning textual data into quantitative signals in this paper. For sample:

$$E = 1, I = 0$$
$$S = 1, N = 0$$
$$F = 1, T = 0$$
$$P = 1, J = 0$$

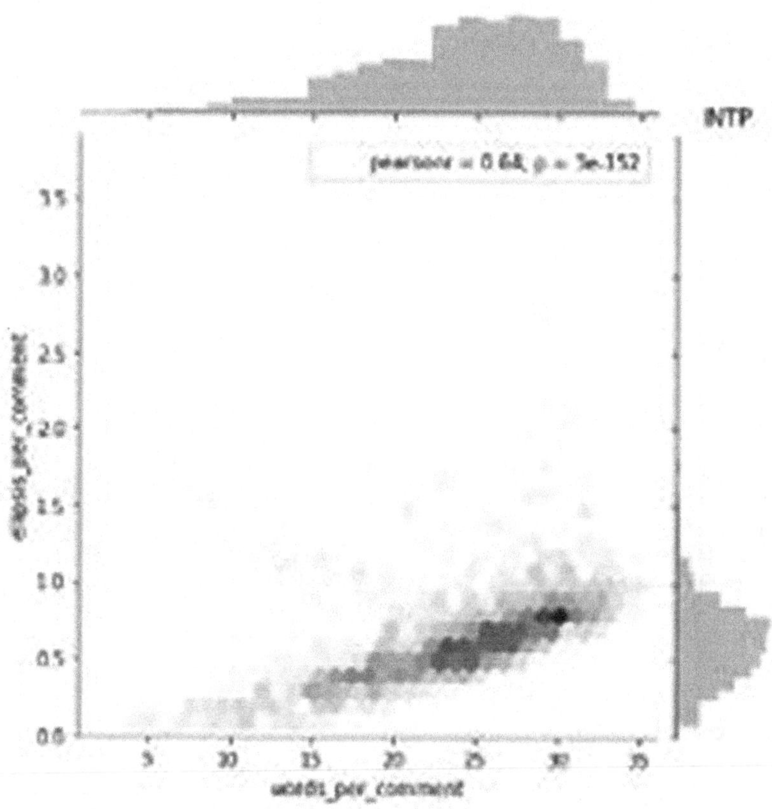

FIGURE 11.12

INTP type.

11.4.1.4 Data split

The dataset was divided into two portions to assess the model's accuracy: the training dataset and the testing dataset. Using the internal module train test split by sci-kit learn. Various splits of the ratios in training and testing where experimented like we used 90%–10%, 80%–2%, 70%–30% and 60%–40% for training and testing respectively with the random state of five. The test dataset is a collection of unknown data which aren't used during training. This is solely used to evaluate the performance of a fully described classifier.

11.4.1.5 Feature extraction

Term Frequency – Inverse Document Frequency (TF-IDF) is used foe feature extraction because it's a prominent tool for text data mining and has promised with good results with the same. It is one of the most essential approaches for representing how relevant a certain word or phrase is to a given document in terms of information retrieval's–idf for feature engineering assesses the relevance/importance of a word to a document in a collection of documents. It is highly important for

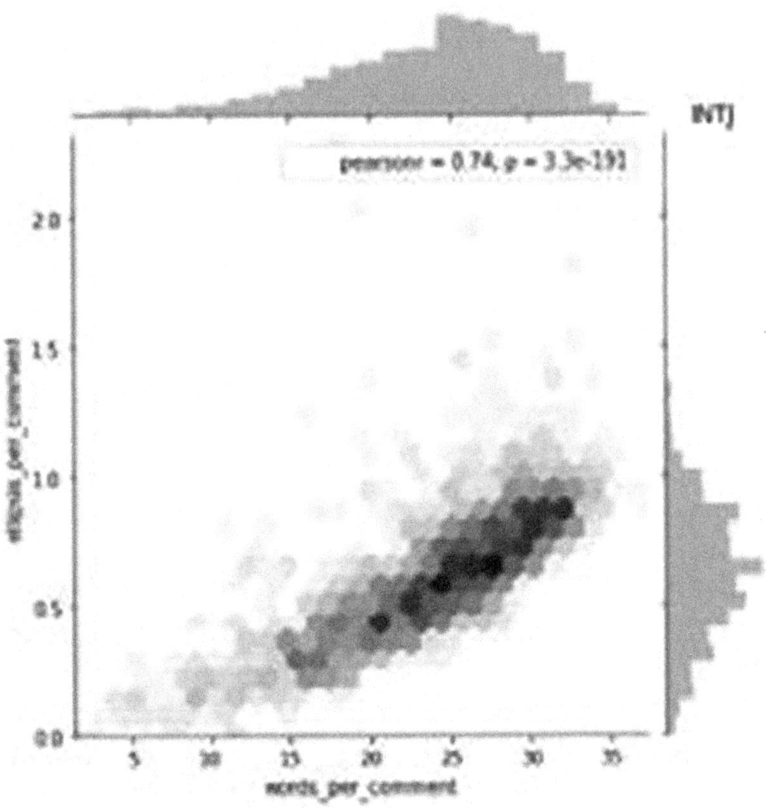

FIGURE 11.13

INTJ type.

scoring words in machine learning methods for Natural Language Processing because we train individual classifiers here. For our model we vectorize using count vectorizer and tf-idf vectorizer keeping the words listing between 10% and 70% of the posts. Count Vectorizer has parameters which are set to analyzer to count, maximum features = 1000, maximum difference = 0.7 and minimum difference = 0.1.

11.4.1.6 Model building for classification

Numpy and sklearn were used to create the Random Forest, Logistic Regression, XG Boost, Stochastic gradient descent, KNN Neighbor and Support Vector Machine (SVM) models. The train test split function was used to divide the data into train and test datasets through the sklearn library. The MBTI type indicators were learned independently during the training phase. The MBTI type indicators were learned independently during the training phase and random state of 7 set constant throughout the experimentation. In total, 90%, 80%, 70% and 60% of the data was used as the train set (data fitting), and 10%, 20%, 30% and 40% was used for test (making a prediction) in

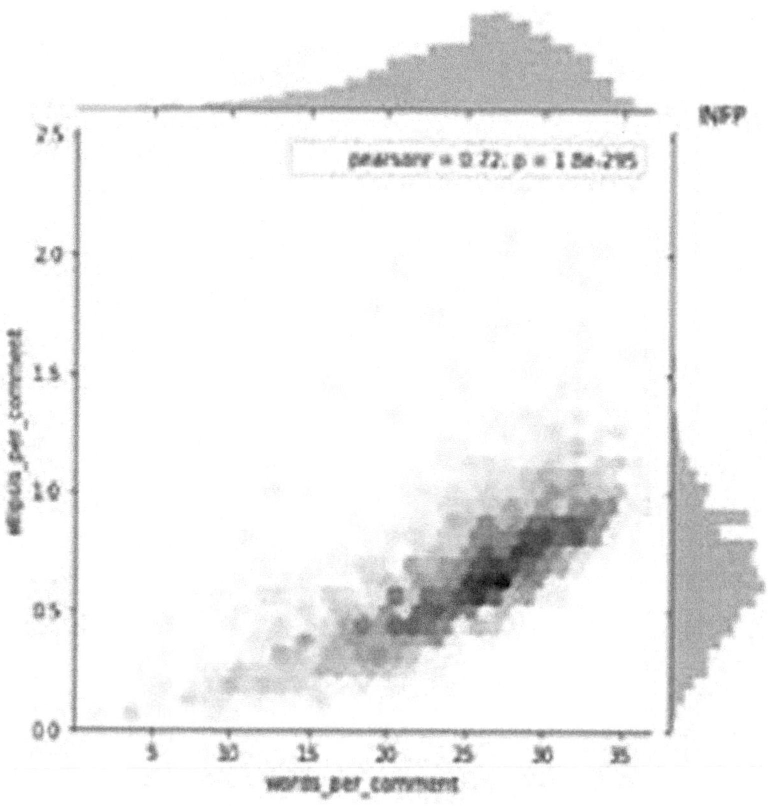

FIGURE 11.14

INFP type.

appropriate training-testing ratio. The hyperparameters set for XGBoost are estimators = 200, max_depth = 2, n_thread(number of threads) = 8 and learning rate = 0.2. The SVM had its random state = 1, throughout the experiment. The process started by removing any columns that aren't relevant to these significance features. Then, using six machine learning methods with the E/I, S/N, T/F and J/P columns was trained and further deep dive was done into the model to obtain a clearer perspective on the prediction. Fig. 11.23 shows the sample output of the classified personality prediction.

The data are derived from the mobile phone through texting and sensors (IOT), is sent to the sever which performs machine learning algorithm, where the Support Vector Machine (SVM) is used. The data received as an input data and the classifier tries to classify the problem based on the emotions and the current psychological mind. Then, the classified output is compared with the personality of the individual and the current support and assistance is provided as per the need. The whole process is then registered in cloud for further assistance and accurate diagnosis by the Psychologist.

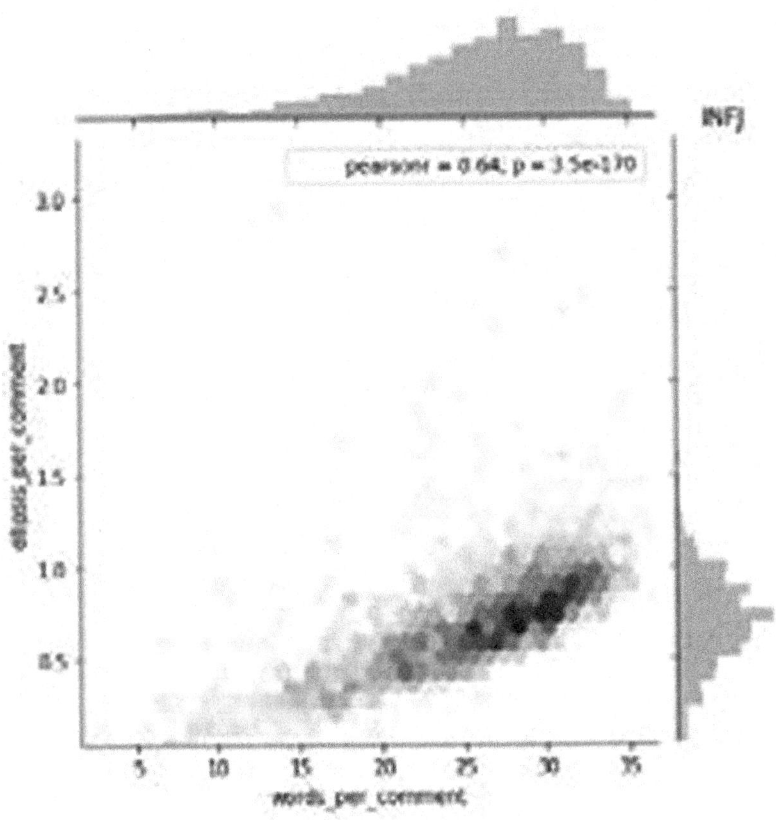

FIGURE 11.15

INFJ type.

11.4.1.7 Comparing accuracy of classification of machine learning models

Using the testing dataset, the accuracy of the Random Forest, Logistic Regression, XG Boost, Stochastic gradient descent, KNN Neighbor and Support Vector Machine (SVM) models, were observed and assessed

11.4.1.8 Evaluating results

The results are evaluated to help discover the best model to describe the data classification of the model. Section 11.4 presents the results of this evaluation. The whole experiment was conducted in Python programming language all the experiments were performed on Google Colaboratory (Colab) with the help of online cloud service and Central Processing Unit (CPU).

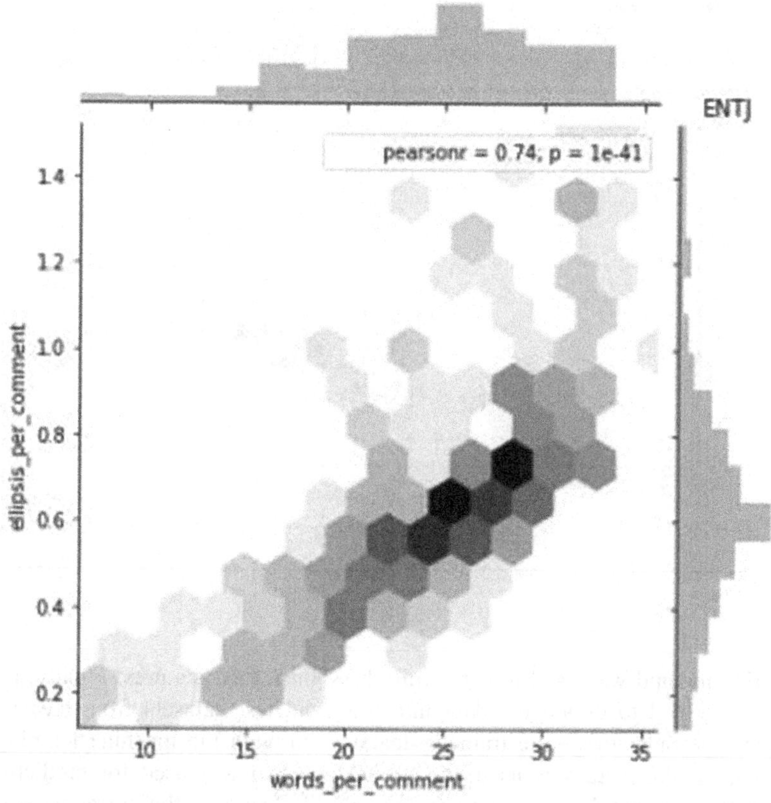

FIGURE 11.16

ENTJ type.

11.5 Results and discussion

The outcomes of the experiment is discussed in this section. Varied number of tests where ran to find the most accurate model for predicting MBTI personality types. To begin, we calculated the number of words each comment in the dataset to identify the MBTI personality types' order. Followed by, increasing the number of characteristics of this experiment because the initial dataset only had two. The Average words per comment and average ellipses per comment are the averages of each attribute was used to analyses these features. The Pearson correlation coefficient was then calculated to determine the strength of the correlations between variables. Since, there is a strong association (69%) between the number of words per comment and the number of ellipses per comment this variable was chosen to trained with various machine learning model.

According to the Pearson correlation, the personality types INFJ, INTP, and ENFP have the highest association between words and ellipses per comment. The dataset was then pre-processed using the word2vec technique to make it more organized and understandable and for feature

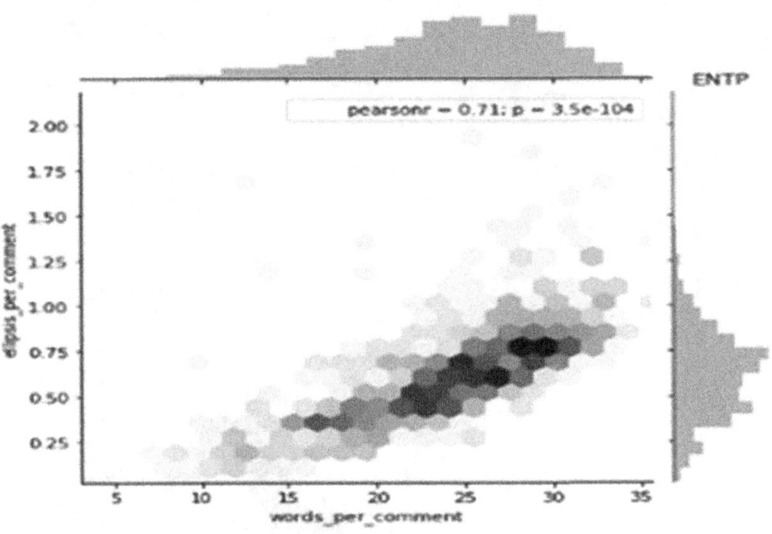

FIGURE 11.17

ENTP type.

extraction TD-IDF method was used in extracting the significance features. Finally, using the train test split function helped to divide the data into train and test datasets from the sklearn library, while the MBTI type indicators were trained separately through the machine learning algorithms. To fit the model, training data was used and the testing data was used for prediction. The final phase entails creating six machine learning models and determining the correctness(accuracies) of each model for every MBTI personality type.

After the completion of classification phase then the model SVM was trained to predict the emotions and psychological mindset through the data received through the IOT devices. The output was compared with the classified personality trait of the individual and proper assistance and support was given. The whole process is then registered in cloud for further assistance and accurate diagnosis by the Psychologist.

In comparison to other machine learning models, all the models comparatively performed at the same base level and does give similar accuracies and trying to different to fit the training model with various ratio is not much of a difference, in all four aspects of MBTI personality types as seen in Table 11.3. For the categories of Intuition(I)/Sensation(S) and Introversion (I)/Extroversion(E), comparatively Stochastic Gradient Descent(SGD) is performing better in different ratios of training and testing dataset for Introversion (I)/Extroversion(E) class than others and K-NN performs worst for Feeling(F)/Thinking(T) classification in all training fits. Over all, the six algorithms perform better and have similar accuracies where Stochastic Gradient Descent comparatively performs better than others and K nearest neighbor is worst epically for F/T class. The results are given in Table 11.3, followed by the comparison of them in a graph of Fig. 11.24. As a result, the Stochastic Gradient Descent outperforms the other six machine learning models on this dataset in

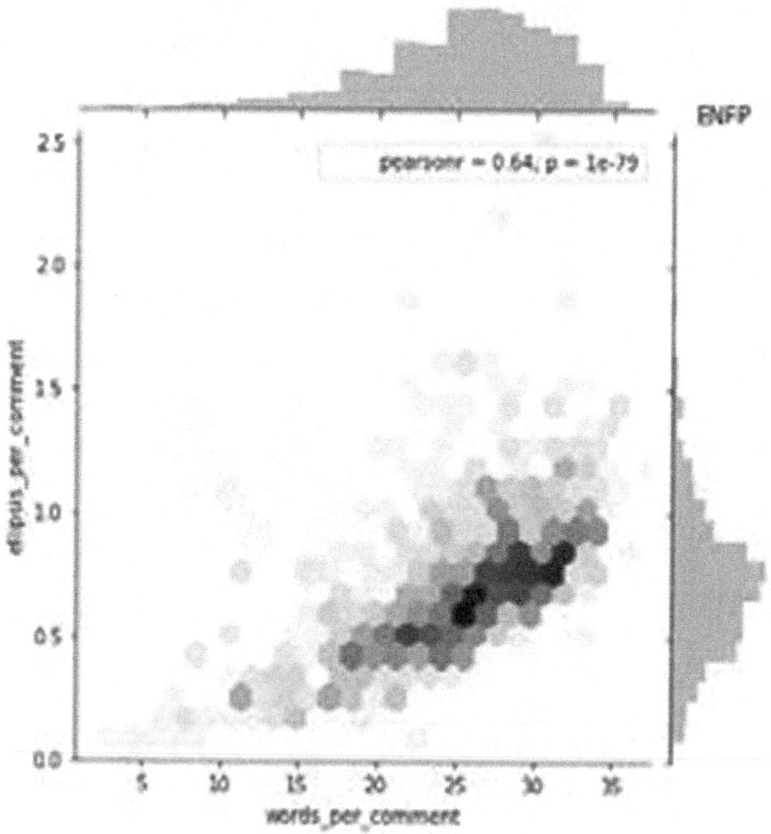

FIGURE 11.18

ENFP type.

terms of overall performance. Performance parameters include Accuracy and Loss for various Training and Testing ratio with all the classifiers. The results are compared against the tabular column Table 11.3 and graph in Fig. 11.24.

This study exclusively looked at those who used a certain social media platform, notably Twitter. Other social media networks may provide useful data, allowing the prediction model to be improved. Furthermore, the aim of this study was solely on the prediction of personality strengths and weaknesses. Apart from that, we must also investigate people's soft skills. Other characteristics, such as thinking and personality, must also be considered.

This study uses a big number of tweets to train the model, and gathering such a vast database for this approach is difficult. This can be improved using a modest large quantity of tweets to review for both training and testing the method's performance. Only English data is used in this work. Focus on many social media platforms or diverse culture scan improve this. Using machine

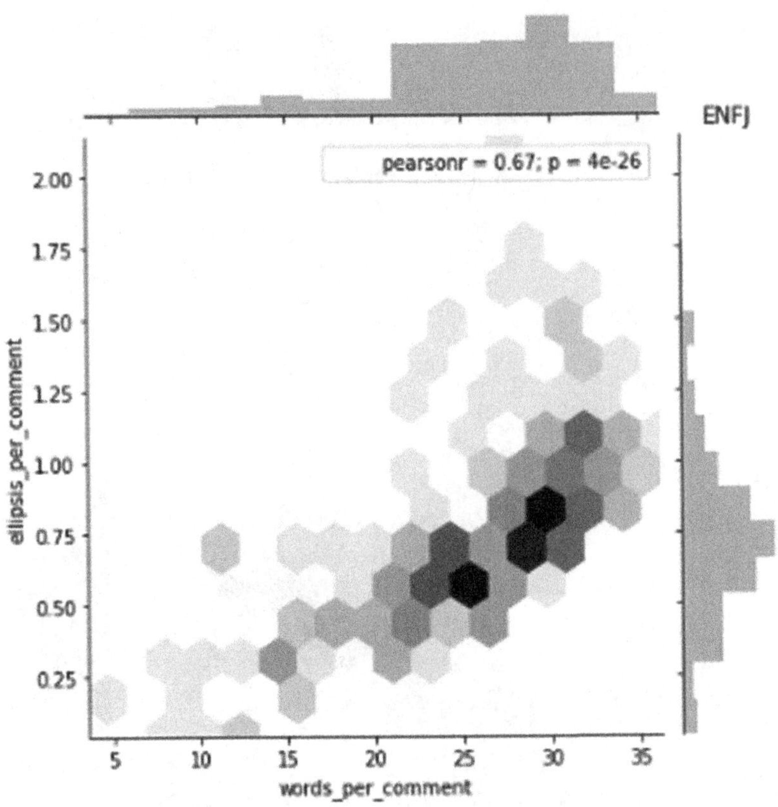

FIGURE 11.19

ENFJ type.

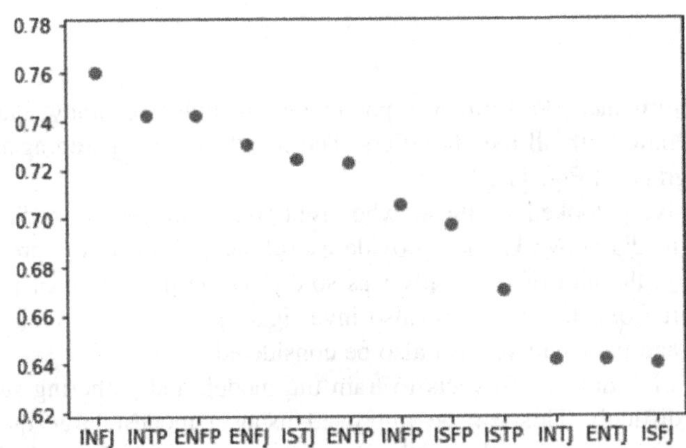

FIGURE 11.20

A pair plot of Pearson Correlation – Ellipses per comment Vs Words per comment.

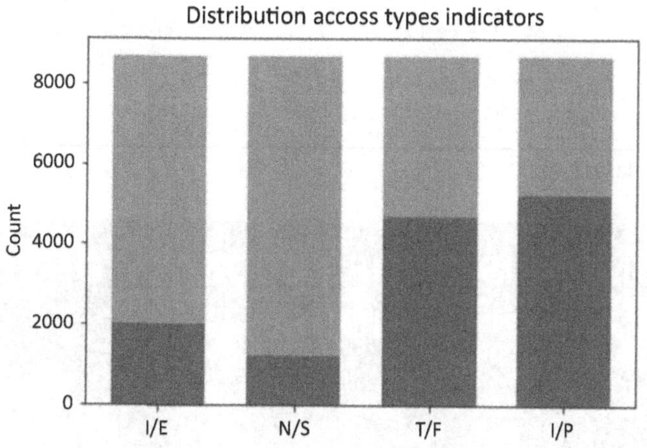

FIGURE 11.21

Distribution of indicators of different types.

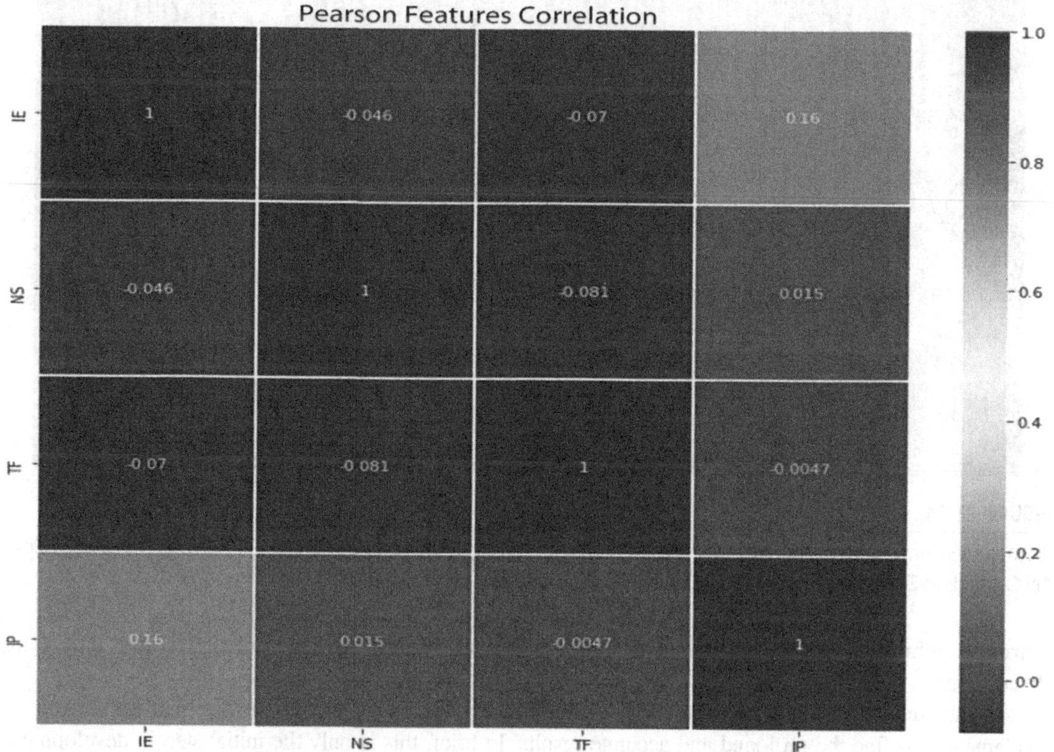

FIGURE 11.22

Pearson correlation coefficient between indicators.

```
[33] print("The result is: ", translate_back(result))

     The result is:  INFJ
```

FIGURE 11.23

Personality prediction classification.

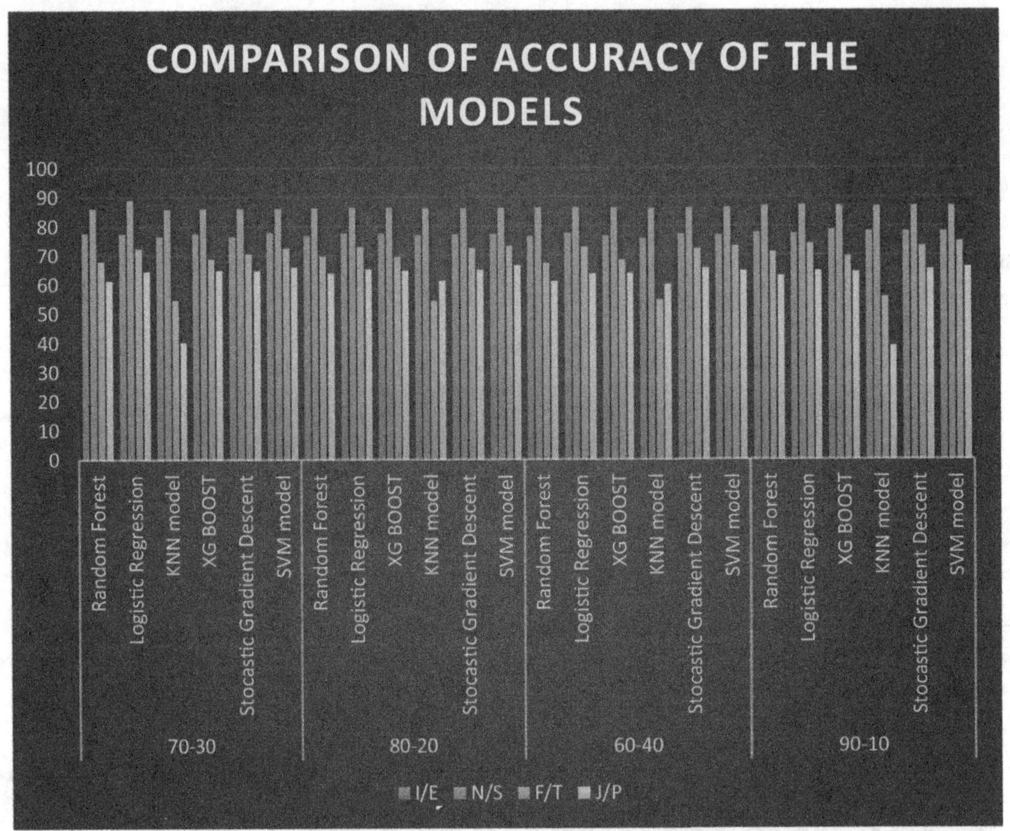

FIGURE 11.24

Comparison of results of different models with varied training and testing ratio. *I/E-Extroversion/Introversion; *N/S-Intuition/Sensation; *F/T-Feeling/Thinking; *J/P-Judgement/Perception.

learning techniques, they may mine a variety of data sources for new insights and intriguing discoveries.

Another scenario is, is where the SVM algorithm is only considered in the later part and yet to be implemented to find the profound and accurate results. In brief, this is only the initial step in developing a model based on personality types on social media comment data and the MBTI personality assessment and an proposed technique with the usage of IOT devices in the field of psychology.

11.6 Conclusion and future works

In this paper, the best model of a machine learning algorithm and utilizing social media data for prediction. Through the study, it's seen that Stochastic Gradient algorithm was is able to predict personality using social media data and also the IOT devices can be used to monitor the individual's condition both physically and mentally for psychological purposes based on personality characteristics. Further, the machine learning algorithm, i.e., SVM can be used to classify the current state to provide the suitable assistance and support in servers. As a result, Psychologists will gain a considerable advantage because they will be able to examine their patient's social media accounts before selecting the best practice. The patients can have a fast and accurate treatments in the absence of the doctor through IoT devices and CPS.

In future works, we intend to gather and generate more datasets in the future to obtain more precise result. We would implement the second phase where IOT devices are used and probably looking forward in adding cloud computing technologies for more accurate and precise results. To improve this prediction system, deep learning algorithm, can be applied as well as their designs and other processes for classification in both the phases. Aside from that, deep learning is a good contender for this problem because it will be able to produce new features from a small collection of features in training datasets, wherein takes less time to analyses large amounts of data.

References

[1] US National Science Foundation, Cyber-Physical Systems (CPS).

[2] O. Hancu, V. Maties, R. Balan, S. Stan, 2007. Mechatronic approach for design and control of a hydraulic 3-dof parallel robot, in: The 18th International DAAAM Symposium, "Intelligent Manufacturing & Automation: Focus on Creativity, Responsibility and Ethics of Engineers."

[3] S.C. Suh, J.N. Carbone, A.E. Eroglu, Applied Cyber-Physical Systems, Springer, 2014.

[4] C.-R. Rad, O. Hancu, I.-A. Takacs, G. Olteanu, Smart monitoring of potato crop: a cyber-physical system architecture model in the field of precision agriculture, Conference Agriculture for Life, Life for Agriculture 6 (2015) 73–79.

[5] S.K. Khaitan, J.D. McCalley, Design techniques and applications of cyber physical systems: a survey, IEEE Syst. J. (2014).

[6] D.Y. Gong, Research on Application of machine learning in data mining, IOP Conf. Ser. Mater. Sci. Eng. 392 (6) (2018).

[7] F. Ahmed, P. Campbell, A. Jaffar, S. Alkobaisi, J. Campbell, Learning & personality types: a case study of a software design course, J. Inf. Technol. Educ. Innov. Pract. 9 (2010) 237–252.

[8] B. de Raad, B. Mlačić, Big five factor model, theory and structure, International Encyclopedia of the Social & Behavioral, second ed., Elsevier, 2015, pp. 559–566.

[9] T.L.C. Yoong, N.R. Ngatirin, Z. Zainol, Personality prediction based on social media using decision tree algorithm, Pertanika J. Sci. Technol. 25 (S4) (2017) 237–248.

[10] N.R. Ngatirin, Z. Zainol, T.L.C. Yoong, A comparative study of different classifiers for automatic personality prediction, in: Proc. – 6th IEEE Int. Conf. Control Syst. Comput. Eng. ICCSCE 2016, 2017, pp. 435–440.

[11] T. Varvel, S.G. Adams, S.J. Pridie, A study of the effect of the myers-briggs type indicator on team effectiveness, ASEE Annu. Conf. Proc. (2003) 9525–9533.

[12] Z. Poursafar, N. Rama Devi, L.L.R. Rodrigues, Evaluation of myers-briggs personality traits in offices and its effects on productivity of employees: an empirical study, Res. Artic. Int. J. Cur Res Rev. 7 (21) (2015) 53–58.

[13] S.D. Mallari, Myers-Briggs Type Indicator (MBTI) Personality Profiling and General Weighted Average (GWA) of Nursing Students, Online Submiss, 2017, pp. 1–11.

[14] A. Furnham, J. Mouttafi, J. Crump, The relationship between the revised NEO-personality inventory and the myers-briggs type indicator, Soc. Behav. Personality 31 (6) (2003) 577–584.

[15] T. Tandera, Hendro, D. Suhartono, R. Wongso, Y.L. Prasetio, Personality prediction system from Facebook users, Procedia Comput. Sci. 116 (2017) 604–611.

[16] W. Rc, Y. Munas, K. Cs, F. Ta, N. Vithana, Personality based e-recruitment system, Int. J. Innov. Res. Comput. Commun. Eng. 5 (2017).

[17] S. Sharma, Predicting employability from user personality using ensemble modelling, 2015, p. 37.

[18] G. Carducci, G. Rizzo, D. Monti, E. Palumbo, M. Morisio, TwitPersonality: computing personality traits from tweets using word embeddings and supervised learning, Information 9 (5) (2018) 1–20.

[19] M.H. Amirhosseini, H. Kazemian, Machine Learning Approach to Personality Type Prediction Based on the Myers – Briggs Type Indicator®, 2020.

[20] N. Dhanda, S.S. Datta, M. Dhanda, Machine Learning Algorithms, no. June 2016, 2019, pp. 210–233.

[21] G.M. Souza, T.A. Catuchi, S.C. Bertolli, R.P. Soratto, Soybean under water deficit: physiological and yield responses, A Comprehensive Survey of International Soybean Research – Genetics, Physiology, Agronomy and Nitrogen Relationships, IntechOpen, 2013.

[22] S. Shalev-Shwartz, Y. Singer, N. Srebro, Pegasos: primal estimated sub-gradient solver for SVM, in: Proceedings of ICML '07.

[23] Y. Tsuruoka, J. Tsujii, S. Ananiadou, Stochastic gradient descent training for l1-regularized log-linear models with cumulative penalty, in: Proceedings of the AFNLP/ACL '09.

[24] X. Wei, Towards optimal one pass large scale learning with averaged stochastic gradient descent.

[25] H. Zou, T. Hastie, Regularization and variable selection via the elastic net, J. R. Stat. Soc. Ser. B 67 (2) (2005) 301–320.

[26] T. Zhang, Solving large scale linear prediction problems using stochastic gradient descent algorithms, in: Proceedings of ICML '04.

[27] Cyber physical systems: NIST engneering labratory, https://www.nist.gov/el/cyber-physical-systems.

[28] Norton: product and services. https://us.norton.com/internetsecurity-iot-what-is-the-internet-of-things.html.

[29] C. Cortes, V.N. Vapnik, Support-vector networks (PDF)Mach. Learn. 20 (3) (1995) 273–297. Available from: https://doi.org/10.1007/BF00994018. CiteSeerX 10.1.1.15.9362. S2CID 206787478.

[30] 1.4. Support Vector Machines—Scikit-learn 0.20.2 documentation." Archived from the original on 2017-11-08. Retrieved 2017-11-08.

[31] T. Hastie, R. Tibshirani, J. Friedman, The elements of statistical learning: data mining, Inference, and Prediction (PDF), second ed., Springer, New York, 2008, p. 134.

[32] B.E. Boser, I.M. Guyon, V.N. Vapnik, A training algorithm for optimal margin classifiers, Proceedings of the Fifth Annual Workshop on Computational Learning Theory – COLT '92, ACM, 1992, p. 144. CiteSeerX 10.1.1.21.3818. Available from: https://doi.org/10.1145/130385.130401. ISBN 978-0897914970. S2CID 207165665.

[33] W.H. Press, S.A. Teukolsky, W.T. Vetterling, B.P. Flannery, Section 16.5. Support vector machines, Numerical Recipes: The Art of Scientific Computing, third ed., Cambridge University Press, New York, 2007. ISBN 978-0-521-88068-8. Archived from the original on 2011-08-11.

[34] MypersonalityCafeDataset: https://www.personalitycafe.com/forums/myers-briggs-forum.49/.

[35] L. Bottou, Stochastic Gradient Descent – Website, 2010.

Further reading

N.H.Z. Abidin, M.A. Remli, N.M. Ali, D.N.E. Phon, N. Yusoff, H.K. Adli, et al., Improving intelligent personality prediction using myers-briggs type indicator and random forest classifier, Int. J. Adv. Computer Sci. Appl. (IJACSA) 11 (11) (2020) Kaggle Dataset. Available from: https://www.kaggle.com/datasnaek/mbti-type.

B. Alouffi, M. Hasnain, A. Alharbi, W. Alosaimi, H. Alyami, M. Ayaz, A systematic literature review on cloud computing security: threats and mitigation strategies, IEEE Access, Digital Object Identifier (2021). Available from: https://doi.org/10.1109/ACCESS.2021.3073203. Received March 17, 2021, accepted April 9, 2021, date of publication April 14, 2021, date of current version April 21, 2021.

An Implementation and Explanation of the Random Forest in Python. [Online]. Available: https://towardsdatascience.com/an-implementation-and-explanation-of-the-random-forest-in-python-77bf308a9b76. (accessed 2 May, 2020).

L. Breiman, Random forests, Mach. Learn. 45 (1) (2001) 5–32.

Chapter 5: Random Forest Classifier – Machine Learning 101 – Medium. [Online]. Available: https://medium.com/machine-learning-101/chapter-5-random-forest-classifier-56dc7425c3e1. (accessed 2 May, 2020).

T. Dai, Y. Dong, Introduction of SVM related theory and its application research, 2020 3rd International Conference on Advanced Electronic Materials, Computers and Software Engineering (AEMCSE), IEEE, 2020, pp. 230–233. Available from: http://doi.org/10.1109/AEMCSE50948.2020.00056.

S. Ghosh, A. Dasgupta, A. Swetapadma, A study on support vector machine based linear and non-linear pattern classification, 2019 International Conference on Intelligent Sustainable Systems (ICISS), IEEE, 2019, pp. 24–28. Available from: http://doi.org/10.1109/ISS1.2019.8908018.

Y. Kumar, K. Kaur, G. Singh, Machine learning aspects and its applications towards different research areas, 2020 International Conference on Computation, Automation and Knowledge Management (ICCAKM), IEEE, 2020, pp. 150–156. Available from: http://doi.org/10.1109/ICCAKM46823.2020.9051502.

F.Q. Lauzon, An introduction to deep learning, 2012 11th International Conference on Information Science, Signal Processing and their Applications (ISSPA), IEEE, 2012, pp. 1438–1439. Available from: http://doi.org/10.1109/ISSPA.2012.6310529.

K. Makwana, G. Dave Dr, A study of identification of personality profiles of undergraduate management students using Myers Briggs type indicator (MBTI) test," (February 1, 2020), Pac. Bus. Rev. Int. 12 (8) (2020).

B.Y. Pratama, R. Sarno, Personality classification based on Twitter text using Naive Bayes, KNN and SVM, Proceeding of 2015 International Conference on Data and Software Engineering ICODSE 2015, IEEE, November 2016, pp. 170–174.

Random Forests Classifiers in Python – DataCamp. [Online]. Available: https://www.datacamp.com/community/tutorials/random-forests-classifier-python#features. (accessed 2 May, 2020).

I.H. Sarker, Machine learning: algorithms, real-world applications and research directions, SN Comput. Sci. 2 (2021) 160.

A. Shrestha, A. Mahmood, Review of deep learning algorithms and architectures, IEEE Access. 7 (2019) 53040–53065. Available from: https://doi.org/10.1109/ACCESS.2019.2912200.

P.D. Tieger, B. Barron-Tieger, Do What You Are: Discover the Perfect Career for You Through the Secrets of Personality Type, fourth ed., Sphere, London, 2007.

D. Upadhyay, J. Manero, M. Zaman, S. Sampalli, Gradient boosting feature selection with machine learning classifiers for intrusion detection on power grids, IEEE Trans. Netw. Serv. Manage. 18 (1) (2021).

CHAPTER

Techniques applied to increase soil fertility in smart agriculture 12

Jyoti B. Deone[1], Rahat Khan Afreen[2] and Viraj R. Jadhao[3]

[1]*Ramrao Adik Institute of Technology Navi Mumbai & BAMU University, Mumbai, Maharashtra, India*
[2]*DIEMS, Bamu University Aurangabad, Maharashtra, India* [3]*Stevens Institute of Technology, NJ, United States*

12.1 Introduction

The demand for organic farming is growing in India with the increase in the popularity of organic products. People have started believing that organic farming is a healthier option and it improves soil fertility. Many farmers have started giving preference to organic farming because of the national and international demand for organic food [1]. In agriculture, soil health is very important. If more chemical fertilizer is added to the soil to increase crop productivity, it may be hazardous for health. It has been proved in scientific studies that crops produced through organic methods taste good and, most importantly, preserve vitamins and minerals, helping to reduce heart ailments, cancer, and many other diseases [2]. Natural fertilizer helps to maintain soil health as well as preserve soil nutrients. Agricultural productivity depends not only on soil nutrients but also on physical, biological, and chemical factors, such as texture, natural content, pH value, and conductivity of soil [3].

While many people are aware of the impact of nuclear technology in the development of electricity, very few know the contribution that radiation has on improving the efficiency and yield of agricultural resources [4]. When we look at the contribution of nuclear energy and technology to the vast domain of agriculture, four primary areas of progress grab attention: bug control, animal wellbeing, higher harvest yield, and enhanced food handling practices [5]. Atomic innovation in expanding crop assembling might be seen in more than one region, but most fundamentally in its impact it has had at the decrease of manure. Numerous ranchers have found addition in diminishing the measure of manure they use in delivering gigantic yields as it decreases charge and reduces any ecological harm. Labeling phenomenal amounts and sorts of manures with radioisotopes allows in for ranchers to immediately relate supplement efficiencies and lacks in light of the fact that the characterized composts are followed as they're ingested into remarkable spaces of the yields [6].

12.2 Related work

Agriculture is a key factor in a country's economic development. In a country like India, agriculture is the primary source of income for the majority of the people [7]. New technologies, such as machine learning and deep learning techniques, are being applied in the agricultural domain to help

Intelligent Edge Computing for Cyber Physical Applications. DOI: https://doi.org/10.1016/B978-0-323-99412-5.00008-3

farmers develop and maximize their output. Natural compost alludes to materials utilized as manure that happen consistently in nature, typically as a side-effect or final result of a normally happening measure [8,9]. Natural composts, for example, excrement have been utilized in agribusiness for millennia; antiquated ranchers didn't comprehend the science in question, however they perceived the advantage of furnishing their yields with natural material [10]. Interest in natural cultivating is becoming worldwide as maintainable horticultural practice these days [11]. Natural composts are supported wellsprings of supplements because of slow delivery during deterioration. By expanding soil natural matter, natural cultivating can reestablish the normal richness of the harmed soil, which will further develop the yield usefulness to take care of the developing populace. Natural composts upgrade the normal soil measures, which have long haul consequences for soil fruitfulness [12,13].

12.3 Different approaches for improving soil fertility

This research study focuses on explaining the following approaches for increasing the fertility of the soil and increasing agricultural productivity.

12.3.1 Precision farming

Agriculture is the backbone of the Indian economy, accounting for 18% of the country's GDP and employing almost half of the workforce. Agriculture provides a living for more than 70% of rural households. With around 17% of the world's total population living in India, the issue of ever-growing population is leading to increasing demand for agricultural supplies [14]. Modernizing conventional agricultural techniques and preparing for a technological revolution is critical in establishing an environment-friendly crop yield system. Smart farming concepts such as precision farming can be implemented sensibly for this purpose [15]. Precision agriculture is an integrated crop management system that uses remote sensors (RS), GPS, and a geographic information system (GIS) to monitor the harvest field close to the ground [16]. The advantage of environmentally friendly and profitable precision farming is evident in the minimal use of water, herbicides, pesticides, and fertilizers in addition to agricultural equipment.

12.3.2 Organic farming

As far as organic farming is concerned, farmers are faced with new challenges and opportunities every day. Organic products are costly because of their small-scale production, as chemicals are not used in their production [17]. Since people are unaware of the authenticity of organic products, farmers are unable to foresee the demand for their products. Farmers are generally unaware of the best organic farming practices as organic farming is not yet widespread [18] in the country. The land for organic cultivation must fulfill some basic requirements, which is, again, difficult to find. Organic farming resources are costly, so farmers are refraining from opting for organic farming to avoid loss. Farmers generally do not opt for organic farming because of financial instability [19].

12.4 Government programs to promote organic farming in India

The arable land for organic farming has increased from 11.83 lakh hectares in 2014 to 29.17 lakh hectares in 2020 because of the government's efforts and new schemes introduced from time to time. The Indian government supports the promotion of organic farming throughout the country through various programs [20]. Some schemes for the encouragement of organic farming [21] adopted by the government of India are discussed below.

1. Paramparagat Krishi Vikas Yojana

   The NDA administration introduced the *paramparagat krishi vikas yojana* (PKVY) in 2015 to encourage organic farming throughout the country. Farmers are urged to form groups or clusters and adopt agricultural practices as part of this initiative, with the goal of forming 10,000 clusters over the next three years. To qualify for the program, each cluster or group must have at least 50 farmers who are ready to begin organic farming under the PKVY and a total area of at least 50 hectares. The government will pay INR 20,000 per acre for three years to each farmer who participates in the scheme.
2. Mission Organic Value Chain Development for North Eastern Region

   The Farmer Producer Organizations (FPOs), encourages third-party certified organic farming of specialist crops, with an emphasis on exports, in the North East area of the country. Farmers receive Rs 25,000 per acre in aid for organic inputs such as organic manure and biofertilizers over a three-year period.
3. Capital Investment Subsidy Scheme (CISS) under Soil Health Management Scheme

   Under this initiative, the state government and government agencies receive 100% aid for the establishment of automated waste compost production units from the fruit and vegetable market, as well as agricultural waste composting units, up to a maximum of Rs 190 lakh per unit (3000 total annual capacity of TPA) Similarly, help is granted to individuals and commercial organizations of up to 33% of the cost ceiling of Rs 63 lakh per unit as a capital investment. Table 12.1 represents data on organic farming in comparison to other nations.

12.4.1 Nuclear technology

One of the most progressive methods of enhancing agricultural practices is nuclear technology. Nuclear programs in agriculture depend on using isotopes and radiation strategies to manipulate

Table 12.1 Data on organic farming in comparison to other nations.

Country	Rank	Organic certification area (in million hectares)
China	3	3.14
USA	7	2.02
India	9	1.94
Brazil	12	1.18

pests and diseases, boost agricultural production, shield land and water resources, ensure the protection and authenticity of meals [22], and boost animal production. Nuclear strategies are currently being used in many nations to assist in holding healthy soils and water delivery systems, which are important for ensuring food safety. For example, in the Republic of Benin, West Africa, a program for 5000 rural farmers multiplied maize yield by 50% and decreased the quantity of fertilizer use by 70% with strategies that facilitate nitrogen fixation. Likewise, nuclear strategies permit Maasai farmers in Kenya to plan small-scale irrigation, doubling vegetable yields, and making use of only 55% of the water that could usually be carried out through the use of conventional guide irrigation [23]. Isotopic and nuclear strategies can play an essential role in presenting precious records for the improvement of techniques to enhance soil fertility and crop performance.

The availability of critical nutrients and water for plants, affected by excess salts in the soil solution, is the fundamental restriction of salty agriculture. Among the basic plant nutrients, N is important for development and production. Nuclear and isotopic methods (sometimes known as nuclear techniques) are a supplement to, not a replacement for, traditional nonnuclear procedures [24]. Nuclear approaches, on the other hand, offer significant benefits over traditional techniques in that they provide unique, accurate, and quantitative data on soil nutrients and soil moisture pools and fluxes in soil, water systems, and the environment. For soil salinity control, isotope methods give helpful information for monitoring soil water nutrient management that may be adjusted to individual agricultural environments [25]. For example, 15 N stable isotope techniques can be used to measure the rates of various N transformation processes in soil, water, and atmospheric systems, such as mineralization, immobilization, nitrification, biological nitrogen fixation, and the sources of production of nitrous oxide (N_2O) in the ground – a gas that contributes to greenhouse effect and is responsible for ozone depletion. The utilization of oxygen-18, hydrogen-2 (deuterium), and other isotopes is essential in agricultural water management since it allows for the identification of water sources as well as the monitoring of water movement and channels. Different irrigation technologies, cropping patterns, and agricultural practices have an impact on water in agricultural landscapes. It also aids in the understanding of plant water usage, measures crop transpiration and soil evaporation, and enables us to develop methods to optimize crop yield, eliminate wasteful water loss, and avoid soil and water degradation.

12.5 Methodology

12.5.1 Dataset collection

The dataset used in this study was created by compiling data from several Government of India sites on weather and soil. The sample of the dataset is shown in Fig. 12.1. These data are relatively simple and have few but useful properties that affect crop yields. The data includes nitrogen, phosphorus, potassium, and soil pH values. It also includes humidity, temperature, and rainfall required for a particular crop.

Soil texture, which determines its ability to retain nutrients, and the degree of acidity or alkalinity (pH level) of the soil are master variables regulating nutrient availability. The activity of microorganisms in the soil, as well as the degree of exchangeable aluminum, is affected by the pH level of the soil. Both water retention and drainage influence root penetration. As a consequence, the criteria listed above are considered while picking a crop for the following reasons: Crop prediction

Out[4]:

	N	P	K	temperature	humidity	ph	rainfall	label
2195	107	34	32	26.774637	66.413269	6.780064	177.774507	coffee
2196	99	15	27	27.417112	56.636362	6.086922	127.924610	coffee
2197	118	33	30	24.131797	67.225123	6.362608	173.322839	coffee
2198	117	32	34	26.272418	52.127394	6.758793	127.175293	coffee
2199	104	18	30	23.603016	60.396475	6.779833	140.937041	coffee

```
In [5]: df.size
Out[5]: 17600

In [6]: df.shape
Out[6]: (2200, 8)
```

FIGURE 12.1

Soil nutrients dataset.

using the ensembling technique. The correlation of nutrients in the soil is shown in Fig. 12.2 with the help of the heat map visualization technique.

The ensemble learning model has been employed in this case. Ensemble learning is a data mining technology, also known as the committee method or model combiner, which combines the strengths of many models to generate better predictions and efficiency than each model alone. In our system, we use majority voting, which is one of the most well-known assembly procedures. The voting method can be carried out with any number of basic students. At least two foundational learners are required. The pupils are chosen in such a way that their skills complement one another. The greater the competition, the better the forecast. However, it is vital for the learners to be complementary since when one or a few members make an error, the possibility of the remaining members rectifying this error is high. Each learner develops into a model. The specified training dataset is used to train the model. When a fresh sample must be categorized, each model predicts the class on their own. Finally, the class predicted by the majority of the learners is chosen as the new sample's class label. The block diagram in Fig. 12.3 represents the overall working of our study and is all about the model built using the algorithms listed below.

1. Decision Tree Algorithm

   It is a classification algorithm, which looks like the structure of a tree, and contains three parts: a node representing an attribute, a branch representing the value of that attribute, and a leaf node representing a class label. It is one of the types of supervised learning algorithms. In this study, we already knew the value of the training data, and it always chose the favorite

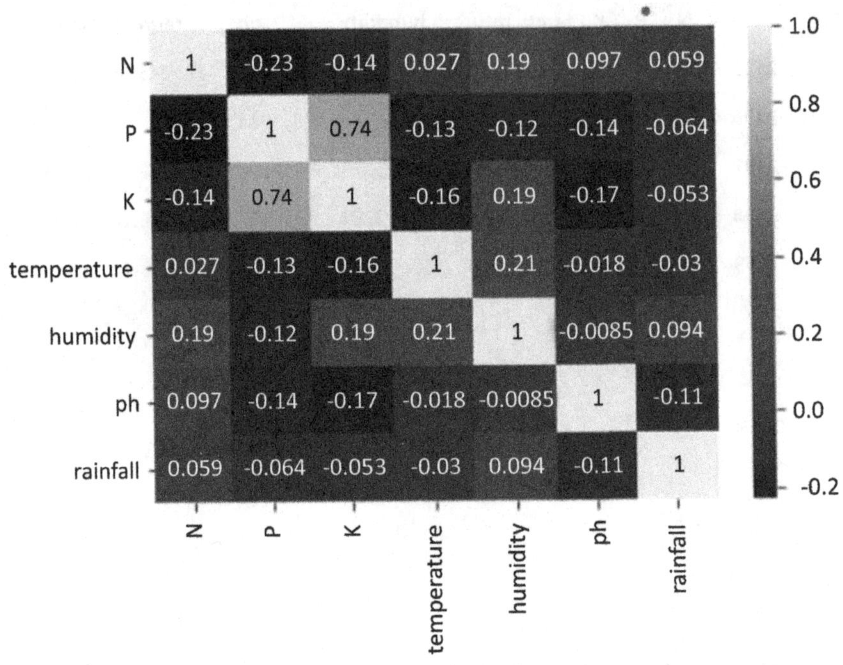

FIGURE 12.2

Correlation between soil nutrients.

attribute as a root node. Decision tree (DT) is a popular algorithm for classification as well as regression. It handles high-dimensional data that can be easily understood and gives good accuracy results [17]. DT is a tracking model that combines a series of core tests in an efficient and consistent manner where a numerical characteristic is relative to a cutoff value in each test. Conceptual rules are much easier to construct than numerical weights in the neural network of connections between nodes. DT is used mainly for grouping purposes. In addition, DT is a classification model generally used in data mining. Each tree is made up of nodes and branches. Each node represents the characteristics of a category to be classified, and each subset defines a value that the node can support. Due to their simple analysis and precision on multiple forms of data, DT algorithms have found many areas for implementation.

2. Naive Bayes Algorithm

The naive Bayes classifier is a probabilistic classifier based on the theory of Bayes. It is based on the idea that the presence of one feature in one class has nothing to do with the presence of another feature in the same class. A watermelon, for example, is a green, spherical fruit around six inches in diameter. A naive Bayes analysis looks at these three qualities, with each contributing to the computation of probability in its own way. For example, we could say that a fruit has a 60% probability of being a watermelon if it is green, a 40% chance of being a watermelon if it is round, and a 95% chance of being a

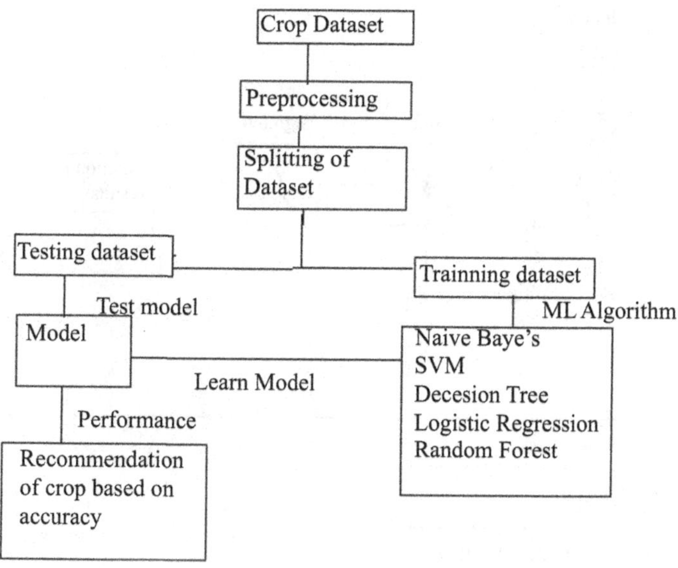

FIGURE 12.3

Overall working model.

watermelon if its diameter is close to 15 cm. It is now referred to as "naive" since all factors independently contribute to the decision-making.

Bayes' theorem is stated as follows Eq. (12.1):

$$P\left(\frac{A}{B}\right) = P\left(\frac{B}{A}\right) \times \frac{P(A)}{P(B)} \tag{12.1}$$

When the probability that we want the calculation of P (AB) is called the rear probability, and the marginal probability of the P (A) event is called before.

3. Support Vector Machine

A support vector machine (SVM) is a powerful and adaptable machine learning technique that can make linear and nonlinear classifications, regression, and outlier identification. Support vector machines, or SVMs, are another approach frequently used in machine learning for classification and regression issues, but they are more commonly employed for classification tasks. It is preferred over other classification algorithms because it requires fewer calculations and provides exceptional precision. The technique is beneficial since it produces trustworthy conclusions even when there is a limited amount of data.

4. Logistic Regression

In Fig. 12.4 the diagram of the whole logistic regression model is shown. The logistic regression model, which is comparable to the SVM, was used in the same way. It establishes unbiased and biased variables. It uses the sigmoid characteristic to forecast possibilities and set decision bounds. The most obvious variation is that for fine-tuning, the l2 penalty and the exceptional regularization parameter values of C were applied.

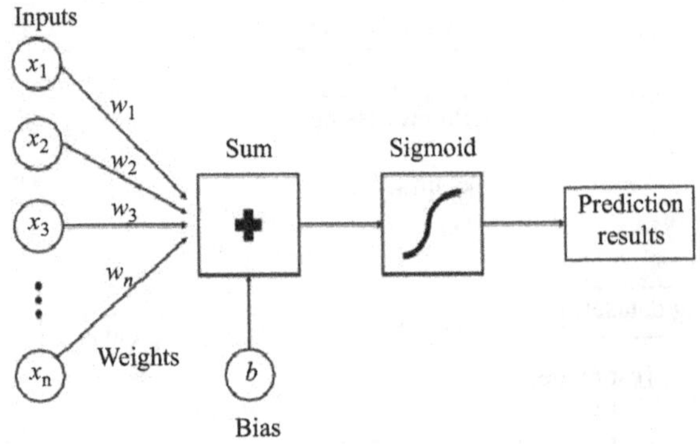

FIGURE 12.4

Working of logistic regression.

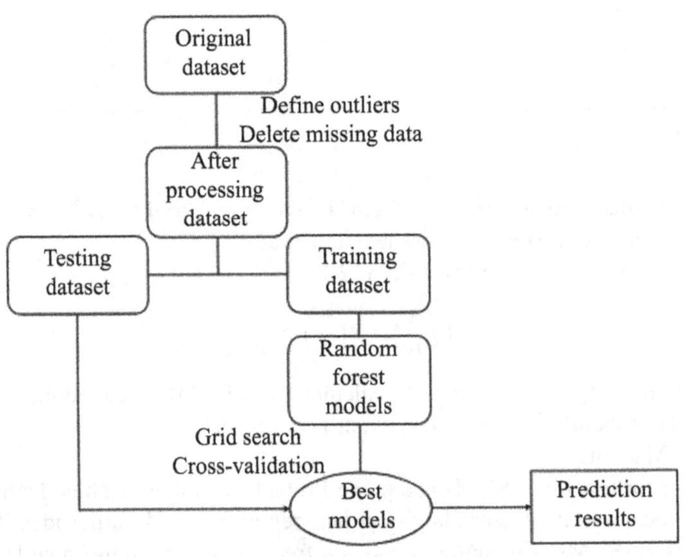

FIGURE 12.5

Working diagram for RF model.

5. Random Forest

Fig. 12.5 shows how the rainforest algorithm works. The approach of this algorithm is very similar to a decision tree. The record is preprocessed, and some samples from the collection of classes are selected. There is a bureaucratic decision tree for each determined sample. The random forest version was initially mastered without tweaking.

Next, similar to SVM, a fivefold cross-validation, and a grid search with a combination of anomalous parameters, such as the number of bushes in a random forest (n estimates) were used. This property should be used for the number of skills. In each case, we should remember the grade divisions in the tree and how to select a sample to train each tree. The Gini standard was used to assess the comfort of wood. Entropy criteria have also been tried in the model, but Gini criteria scores are more accurate.

12.6 Model comparision and evaluation

After applying the given dataset on all models, we observed that the random forest model has greater accuracy compared with all other models, as shown in Fig. 12.6. We used a random forest algorithm to get the correct crop recommendation (Table 12.2).

When input values from our dataset are provided we get the correct crop as expected, as shown in Fig. 12.7.

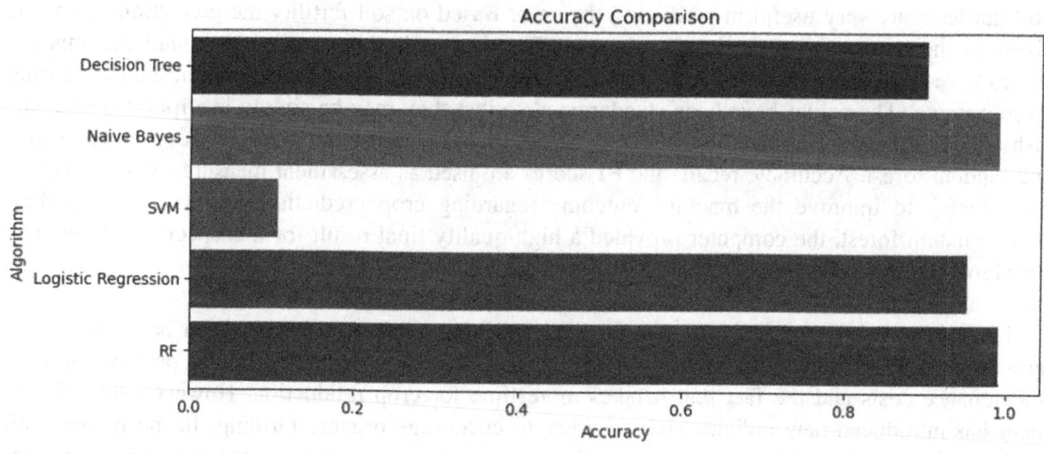

FIGURE 12.6

Accuracy comparasion.

Table 12.2 Accuracy comparison.

Model	Decision tree	Naive Bayes	Support vector machine	Logistic regression	Random forest (RF)
Accuracy	0.96	0.99	0.97	0.95	0.99

Making a prediction

```
In [43]:
data = np.array([[104,18, 30, 23.603016, 60.3, 6.7, 140.91]])
prediction = RF.predict(data)
print(prediction)

['coffee']

In [44]:
data = np.array([[83, 45, 60, 28, 70.3, 7.0, 150.9]])
prediction = RF.predict(data)
print(prediction)

['jute']
```

FIGURE 12.7

Prediction of crop using RF model.

12.7 Conclusion

Soil nutrients are very useful in predicting the crop. Based on soil fertility the prediction of crop is given as the "crop recommendation" dataset. The dataset has been displayed, and the missing values have been squeezed into it. A few superfluous attributes were removed from the data during preprocessing. The values have been standardized so that they may be simply incorporated into ML fashions. The dataset was then used to train SVM, logistic regression, decision trees, naive Bayes, and random forest. Accuracy, recall, and F1 scores are used as assessment measures. Our focus was on accuracy to improve the machine outcome regarding crop prediction based on soil fertility. Using random forest, the computer provided a high-quality final result for this specific dataset. The RF algorithm was used for the prediction of crops based on soil fertility to achieve the best results across all modes.

The study also focuses on precision farming and how nuclear technology can be useful in the agriculture sector. Organic farming is not very popular among Indian farmers because of high maintenance costs and the fact that it takes more time for crop production. However, the government has introduced new policies and subsidies to encourage organic farming. In the future, with the help of machine learning, it will be possible to predict good fertilizer according to crop requirements, as well as how to improve soil quality and crop quality, using a larger dataset and more machine learning models.

References

[1] J. Alzubi, A. Nayyar, A. Kumar, Machine learning from theory to algorithms: an overview, J. Phys. Conf. Ser. 1142 (2018). Article ID 012012.

[2] E. Thippeswamy, Comparative analysis of organic and inorganic food, IOSR J. Agric. Vet. Sci. 4 (6) (2013) 53–57. Available from: https://doi.org/10.9790/2380-0465357.

[3] L. Wu, Y. Jiang, F. Zhao, X. He, H. Liu, K. Yu, Increased organic fertilizer application and reduced chemical fertilizer application affect the soil properties and bacterial communities of grape rhizosphere soil, Sci. Rep. (2020). Available from: https://doi.org/10.1038/s41598-020-66648-9.

[4] M.M. Selim, Introduction to the integrated nutrient management strategies and their contribution to yield and soil properties, Hindawi Int. J. Agron. 2020 (2020). Available from: https://doi.org/10.1155/2020/2821678.

[5] A. Priyadharshini, S. Chakraborty, A. Kumar, O.R. Pooniwala, Intelligent crop recommendation system using machine learning, Proceedings – 5th International Conference on Computing Methodologies and Communication, ICCMC 2021, IEEE, 2021, pp. 843–848. Available from: http://doi.org/10.1109/ICCMC51019.2021.9418375.

[6] A.I. Id, et al., Manure combined with chemical fertilizer increases rice productivity by improving soil health, post-anthesis biomass yield, and nitrogen metabolism, PloS One (2020). Available from: https://doi.org/10.1371/journal.pone.0238934.

[7] A. Adamou, A. Bationo, R. Tabo, S. Koala, Improving soil fertility through the use of organic and inorganic plant nutrient and crop rotation in Niger, 2007.

[8] 2018 3rd International Conference on Computational Systems and Information Technology for Sustainable Solutions (CSITSS). IEEE, 2018.

[9] D. Bhardwaj, M.W. Ansari, R.K. Sahoo, N. Tuteja, Biofertilizers function as key player in sustainable agriculture by improving soil fertility, plant tolerance and crop productivity, Microb. Cell Factories 13 (1) (2014) 1–10. Available from: https://doi.org/10.1186/1475-2859-13-66.

[10] B. Chinthapalli, A comparative study on the effect of organic and inorganic fertilizers on agronomic performance of faba bean (*Vicia faba* L.) and pea (*Pisum sativum* L.), Agric. For. Fish. 4 (6) (2015) 263. Available from: https://doi.org/10.11648/j.aff.20150406.15.

[11] A.D. Chitimus, et al., Influence of soil fertilization systems and crop rotation on physical and chemical properties of the soil, in: 2020 7th International Conference on Energy Efficiency and Agricultural Engineering (EE&AE) 2020 – Proceedings, November 2020. Available from: https://doi.org/10.1109/EEAE49144.2020.9278974.

[12] M.A. Badawi, S.E. Seadh, E.S.B. Naeem, A.S.E.I. El-Iraqi, Effect of phosphorus fertilizer levels on productivity and grains quality of some rice cultivars, J. Plant Prod. 8 (2017) 411–415. Available from: https://doi.org/10.21608/jpp.2017.39996.

[13] I.H. Sarker, Machine learning: algorithms, real-world applications and research directions, SN Comput. Sci. 2 (2021) 160. Available from: https://doi.org/10.1007/s42979-021-00592-x.

[14] H. Kendall, B. Clark, W. Li, et al., Precision agriculture technology adoption: a qualitative study of small-scale commercial "family farms" located in the North China Plain, Precis. Agric. (2021). Available from: https://doi.org/10.1007/s11119-021-09839-2.

[15] R. Finger, S.M. Swinton, N. El Benni, A. Walter, Precision farming at the nexus of agricultural production and the environment, Annu. Rev. Resour. Econ. 11 (2019) 313–335. Available from: https://doi.org/10.1146/annurev-resource-100518-093929.

[16] S.M. Say, M. Keskin, M. Sehri, Y.E. Sekerli, Adoption of precision agriculture technologies in developed and developing countries, Online J. Sci. Technol. 8 (2018) 7–15.

[17] D. Jyoti Bhanudas, K. Rahat Afreen, Prediction of soil accuracy using data mining techniques, 2019 5th International Conference On Computing, Communication, Control and Automation (ICCUBEA), IEEE, 2019, pp. 1–5. Available from: http://doi.org/10.1109/ICCUBEA47591.2019.9129579.

[18] A. Le Campion, F.X. Oury, E. Heumez, et al., Conventional vs organic farming systems: dissecting comparisons to improve cereal organic breeding strategies, Org. Agr. 10 (2020) 63–74. Available from: https://doi.org/10.1007/s13165-019-00249-3.

[19] G.T. Patle, S.N. Kharpude, P.P. Dabral, V. Kumar, Impact of organic farming on sustainable agriculture system and marketing potential: a review, Int. J. Environ. Clim. Change 10 (11) (2020) 100–120. Available from: https://doi.org/10.9734/ijecc/2020/v10i1130270.

[20] A. Kaşif, G. Ortaç, E. İbiş, T.T. Bilgin, Performing similarity analysis on organic farming crop data of Turkish Cities, 2020 Innovations in Intelligent Systems and Applications Conference (ASYU), IEEE, 2020, pp. 1–4. Available from: http://doi.org/10.1109/ASYU50717.2020.9259831.

[21] S. Adams Inkoom, Encouraging Organic Agriculture: The Effects of Conversion Subsidies, Sourthern University, 2017.

[22] D. Fonseca-López, N.J. Vivas Quila, H.E. Balaguera-López, Techniques applied in agricultural research to quantify nitrogen fixation: a systematic review, Manage. Environ. Sustain. (2020). Available from: https://doi.org/10.21930/rcta.vol21_num1_art:1342.

[23] M. Zaman, S.A. Shahid, L. Heng, The role of nuclear techniques in biosaline agriculture, Guideline for Salinity Assessment, Mitigation and Adaptation Using Nuclear and Related Techniques, Springer, Cham, 2018. Available from: https://doi.org/10.1007/978-3-319-96190-3_6.

[24] B.W. Brook, A. Alonso, D.A. Meneley, J. Misak, T. Blees, J.B. van Erp, Why nuclear energy is sustainable and has to be part of the energy mix, Sustain. Mater. Technol. 1–2 (2014) 8–16. Available from: https://doi.org/10.1016/j.susmat.2014.11.001. ISSN 2214–9937.

[25] K. Kubo, H. Kobayashi, M. Nitta, et al., Variations in radioactive cesium accumulation in wheat germplasm from fields affected by the 2011 Fukushima nuclear power plant accident, Sci. Rep. 10 (2020) 3744. Available from: https://doi.org/10.1038/s41598-020-60716-w.

Index

Note: Page numbers followed by "*f*" and "*t*" refer to figures and tables, respectively.

D

E

F

K

L

M

Q

R

S

Printed in the United States
by Baker & Taylor Publisher Services